The Gross Physiology of the Cardiovascular System

Robert M. Anderson, M.D.

Emeritus Associate Dean and Associate Professor of Surgery,
University of Arizona College of Medicine

Former Chief of Cardiothoracic Surgery,
University of Arizona Medical Center

Fellow of the American College of Cardiology

Fellow of the American College of Surgeons

Diplomate, American Board of Surgery

This edition Copyright © 2012 by the estate of Robert M. Anderson, MD. This text is made available under a Creative Commons Attribution-NonCommercial 3.0 License. It may be copied, distributed, excerpted, or adapted for noncommercial purposes so long as proper attribution is made to this original work.

ISBN 978-1-105-52223-9

This text is provided for informational and educational use only, on an as-is basis. Nothing contained herein is or should be considered, or used as a substitute for, medical advice, diagnosis, or treatment by professional healthcare providers. The parties involved in the preparation or publication of the text do not represent or warrant the accuracy, completeness, correctness, usefulness, or timeliness of any information contained herein. They make no representations or warranties with respect to any treatment, action, or application of medication by any person following the information provided herein, and will not be liable for any direct, indirect, consequential, special, exemplary, or other damages arising therefrom.

Citation for the original 1993 print version:

Anderson, Robert M. *The Gross Physiology of the Cardiovascular System.* Tucson, AZ: Racquet Press, 1993. Print.
  ISBN: 0961752815
  ISBN-13: 9780961752811

This text and other materials are available online at:
http://www.cardiac-output.info

Table of Contents

Introduction — 1

Chapter 1: Normal Circulation — 3

Chapter 2: Abnormal Circulation — 15

Chapter 3: Open Heart Surgery with Passive-Filling Pumps — 22

Chapter 4: Animal Experiments with Passive-Filling Pumps — 31

Chapter 5: Hydraulic Model of the Cardiovascular System — 47

Summary — 60

Appendix: Clinical Measurement of the Mean Cardiovascular Pressure — 61

About the Author — 63

Additional Resources — 64

Introduction

At a time when knowledge about microvascular physiology and subcellular myocardial and vascular biochemistry has accumulated at such a tremendous rate, I perceive that a realistic global understanding of the cardiovascular system has been partially lost in that voluminous accumulation of minutiae. In order to be able to see "the forest, and not just individual leaves on the trees," one must first have a clear understanding of the gross mechanical function of the cardiovascular system as a whole.

In determining whether to replace older "sacred cows" with newer concepts, your litmus test should be the extent to which each explains observations in both normal and pathological states. Thus, a valid concept of cardiovascular physiology must be compatible with the following frequently overlooked facts:

- A blood volume equilibrium persists between the systemic and pulmonary circuits even when there are massive shunts between the two, and remains even after closure of those shunts.

- During cardiopulmonary bypass, the empty heart may continue to beat strongly even in the absence of any diastolic filling or stretching of the ventricles.

- Booster pumping does not increase circulation rate in the absence of heart failure.

- Increasing pacemaker rate above that necessary to prevent failure does not increase cardiac output.

- Ventricular pressures measured during heart catheterization are always above zero (in relation to the intrathoracic pressure).

- After heart transplantation, without nerve supply to the heart or artificial pacing, the cardiac output and pulmonary/systemic blood volume balance remain normal.

- In the absence of heart failure, an increase in arterial resistance does not reduce cardiac output.

An overall concept of cardiovascular physiology should accommodate these facts and all other available data. The concept presented in the following chapters explains and accommodates all of these findings.

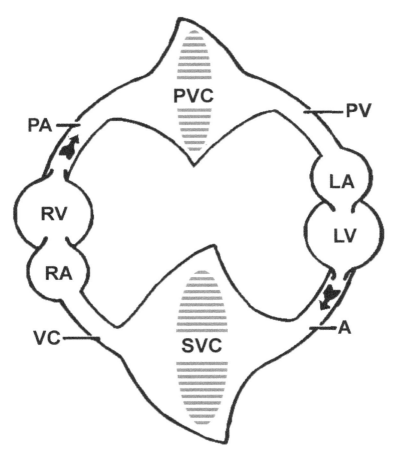

Fig. 1. THE ELASTIC, FLUID FILLED, CARDIOVASCULAR CIRCLE. Diastolic blood volume in the left ventricle (LV) is pumped into the aorta (A), from which it flows sequentially into the systemic vascular compartment (SVC), vena cavae (VC), right atrium (RA), right ventricle (RV), pulmonary artery (PA), pulmonary vascular compartment (PVC), pulmonary veins (PV), left atrium (LA), left ventricle (LV), "and around and around it goes." As the ventricles can only pump out the blood that runs into them, you might ask "Which came first, the chicken or the egg?" We shall see.

Chapter 1:
Normal Circulation

The cardiovascular system has ten unique characteristics that make it an unusually complicated hydraulic system. Understanding how the cardiovascular system functions requires insight into a larger set of variables than that which governs the function of most pump, pipe, and fluid systems found in the world of man-made machines. The ten unique characteristics peculiar to the cardiovascular system are:

1. The system is a closed circle rather than being open-ended and linear.

2. The system is elastic rather than rigid.

3. The system is filled with liquid at a positive mean pressure ("mean cardiovascular pressure"), which exists independent of the pumping action of the heart.

4. The right and left ventricles, which pump into the same system that they pump out of, are in series with two interposed vascular beds (systemic and pulmonary).

5. The heart fills passively, rather than by actively sucking.

6. As a consequence of the heart's passive filling, the circulation rate is normally regulated by peripheral-vascular factors, rather than by cardiac variables.

7. The flow from the heart is intermittent, while the flow to it is continuous.

8. Normally, there is an excess expenditure of energy by the heart needed for the circulation rate imposed by peripheral vascular regulators ("pump energy excess").

9. Normally, ventricular capacity is in excess of the diastolic filling volume ("pump capacity excess").

10. The slowing effect of any vascular resistance on flow rate depends on its location, with reference to upstream compliance, as well as its magnitude.

In order to understand circulatory phenomena in the elastic circular hydraulic system (Fig. 1), where every point is both upstream and downstream from every other point, and where the non-sucking ventricles pump out of the same system they pump into, we need to examine the ten unique factors individually, before we can amalgamate them into a meaningful whole.

Ventricles (The Pumps)

To clarify and facilitate understanding of the features peculiar to the heart, it is helpful to compare three major types of pumps. The heart's unique characteristics as a pump are of paramount importance in understanding how the cardiovascular system works.

PUMP TYPE #1: This type of pump both sucks and forcibly ejects fluid. This pump uses energy both to actively fill at its inlet and to empty its contents at its outlet. Two examples of this type of pump are: (1) a piston pump, which expends energy to suck in the stroke volume which is then forcibly ejected; and (2) a roller pump, which sucks at its inlet by the recoil of the resilient tubing that has been compressed by the roller as it moves forward, ejecting the fluid in front of it. With type #1 pumps, the output in a hydraulic system is determined exclusively by two pump variables: the stroke rate and the stroke volume. The reason for discussing this is that the heart is a different type of pump, and those two variables are frequently and erroneously projected onto cardiovascular function, in a way in which they do not apply.

PUMP TYPE #2: This pump sucks and blows but, instead of producing a specific flow rate, creates a specific pressure gradient between its inlet and outlet. Centrifugal pumps fall into this category. With this type, two pump factors (rate and power), as well as two non-pump factors (pressure and resistance in the system), effect output. As the heart is not such a pump, there is danger in borrowing explanations of cardiovascular function from hydraulic systems containing centrifugal pumps.

PUMP TYPE #3: This type of pump is passive filling, and does not suck at its inlet. It expends no energy to fill, it only expends energy to empty. An example of this type of pump is the urinary bladder. It is a flaccid, hollow organ that does not create any negative pressure or suck on the ureters or kidneys to fill. The bladder merely exerts energy to empty. To calculate the flow of urine for any given period, you can obtain the answer by multiplying the stroke rate times the stroke volume. However, it is important to underline here that those two things, stroke rate and volume, are not determinants of bladder output. You cannot increase the output merely by changing the rate of bladder contraction (stroke rate). The urinary bladder cannot expend energy to increase its filling and thus its stroke volume. This type of pump, even though it does the work and thus produces the flow, is totally dependent upon external factors (e.g., renal function) to determine the output. At a given rate of urinary production, stroke rate and stroke volume are reciprocals of one another. If the bladder is emptied twice as often, the stroke volume will be one half as much.

The Right and Left Ventricles Are Two Type #3 Pumps

The heart, like the urinary bladder, is a hollow muscular organ that does not suck to fill, but produces circulation by ejecting whatever fluid enters it at diastole. During normal function, the heart not only doesn't develop a pressure negative to the intrathoracic pressure, but it offers an impediment to filling because of its limited volume-pressure compliance.

The evidence that the heart fills passively, and does not suck to fill, is found in the data from innumerable heart catheterizations, which all show a positive diastolic pressure in the ventricles (Fig. 2, a to c). In fact, as ventricles fill, they not only do not suck, but they offer an increasing impediment to filling, as noted by the progressive increase in pressure toward the end of diastole (Fig. 2, c). Negative ventricular pressure, in relation to intrathoracic pressure, has not been found in physiologic states. The heart, like all other muscles in the body, expends energy and does work by contracting. It cannot expend energy to do mechanical work by forcibly elongating its fibers to suck like the type #1 pump, which uses energy to suck in a stroke volume. Do not confuse the negative pressure created in the chest by inspiration as negative heart pressure (Fig. 2, b). The chest can suck, the heart cannot.

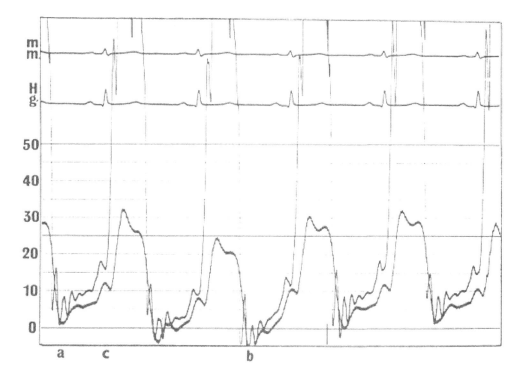

Fig. 2. Right and left ventricular pressures in a normal adult with a closed chest during (a) expiration, (b) inspiration and (c) end diastole.

One can calculate cardiac output in the same way as in the urinary bladder example: by multiplying stroke rate times stroke volume. Also, just as in that example, while stroke rate multiplied by stroke volume measures the amount of output, those variables are not determinants of that output. The heart, by filling passively, pumps out blood at a rate determined by the rate of blood coming to it. Given that the heart is a pump that produces the flow but has a flow rate determined by extra-cardiac factors, let us examine those factors that determine circulation rate.

Determination of Cardiac Output in a Circular Elastic System with Passive-Filling Pumps

With pumps that cannot suck to fill, there must be a positive pressure at the inlets for any blood to run into the ventricles, in order for there to be any pump output. If there is no pressure in the cardiovascular system, no blood can run into the ventricles and there can be no flow. Normally, there is a mean cardiovascular pressure above zero, which the heart distributes. The heart, rather than being responsible for the pressure in the vascular system, is a circulating device. It lowers the pressure at the ventricular inlets and raises it at the ventricular outlets. With a positive pressure in the cardiovascular system, when blood is ejected into the arterial side of the circle, a pressure gradient is created between the arteries and veins. This gradient causes blood to flow around the circle back to the ventricular inlets. Therefore, the output rate varies directly with the magnitude of that mean cardiovascular pressure. <u>The higher the pressure, the higher the gradient, the greater the flow rate</u>. The circular system being elastic, and having resistance and other impediments to flow, the energy from ventricular contraction does not transfer instantaneously around the circle after each heart beat, as would occur in a rigid system. The energy of venous flow is several heart beats behind that of ventricular ejection. <u>The greater the elasticity and impediments to flow, the slower the flow rate.</u>

<u>Therefore, during normal function, cardiac output varies directly with the mean cardiovascular pressure and inversely with the impedance to blood flow to the heart.</u>

Mean Cardiovascular Pressure

Definition: The mean cardiovascular pressure is the pressure in the cardiovascular system with the circulation stopped, after the pressure has equalized between the arteries, capillaries, veins, and cardiac chambers. Do not confuse this pressure with central venous pressure, venous filling pressure, or mean arterial pressure. Mean cardiovascular pressure is the pressure related to the blood volume and the compliance of the entire elastic cardiovascular compartment.

Measurement: Mean cardiovascular pressure is expressed in centimeters of water above ambient pressure, with zero being at mid-heart level. The mean cardiovascular pressure can be approximated during cardiac arrest. After arrest, a pressure equalization occurs between the various cardiovascular compartments in about thirty seconds. The arterial pressure falls and the venous pressure rises as some of the arterial blood moves into the veins during pressure equalization. Therefore, the mean cardiovascular pressure is always above venous pressure and below arterial pressure. Normally, mean cardiovascular pressure is between 15 and 18 cm. of water above mid-heart level. We have some approximation of its magnitude from fortuitous records of the arterial and venous pressures equalizing, obtained during short periods of cardiac arrest of patients in coronary care settings, in emergency rooms, or in operating rooms during heart surgery. Even in these situations, the resulting pressure can be regarded as only an approximation, as the shift in fluid, from hypoxia caused by lack of circulation, may have altered the vascular compliance. (See *Appendix* for clinical measurement technique.)

Significance: Without mean cardiovascular pressure there would be no circulation. The heart doesn't generate the pressure in the vascular system, it merely distributes the mean cardiovascular pressure. The cardiac ventricles take the mean cardiovascular pressure and distribute it by raising the pressure on the arterial sides while lowering it on the venous sides. The two ventricles, being passively filling pumps, cannot suck, so they lower the inlet pressures toward — but never below — zero, in relation to the ambient pressure in the chest.

The higher the mean cardiovascular pressure, if the ventricles are not failing, the higher the ventricles can elevate the arterial pressure while reducing the venous pressure toward zero, and thus the greater the cardiac output. Conversely, the lower the mean cardiovascular pressure, the less the heart can raise the arterial pressure by lowering the venous pressure, and thus the lower the cardiac output.

Origin of the mean cardiovascular pressure: The mean cardiovascular pressure results from the volume of blood and the compliance of the cardiovascular system. The volume in the cardiovascular system results from an equilibrium between the rate of water, electrolytes, and other blood constituents entering the body by way of the gastrointestinal tract, and leaving the body primarily by the kidneys (Fig. 3). The mean cardiovascular pressure is the result of a continuing dynamic process.

**MEAN CARDIOVASCULAR PRESSURE =
energy forcing fluid into the body / resistance to fluid loss from the body**

Homeostatic maintenance of normal mean cardiovascular pressure:

(1) Slow feedback mechanism:

A slow homeostatic feedback mechanism tends to keep the mean cardiovascular pressure at a constant level: Elevation of the mean cardiovascular pressure above normal —› causes increase in cardiac output —› causes increased renal blood flow —› results in increased renal output —› thereby lowering blood volume and mean cardiovascular pressure back to normal. Conversely, low mean cardiovascular pressure —› low cardiac output —› low renal blood flow —› decreased renal output until the mean cardiovascular pressure is restored to normal by continuing fluid intake. With elevated mean cardiovascular pressure, the rate of return to normal is dependent on renal function, whereas, with low mean cardiovascular pressure the rate of return can vary greatly, depending on the rate of restoration of blood volume.

Clinical evidence of the homeostatic mechanism:

- Response to weightlessness by going into orbit: In the absence of gravity, normal blood volume shifts centrally from the lower part of the body, thereby, increasing the mean cardiovascular pressure at heart level —› resulting in increased cardiac output —› causing greater renal blood flow —› leading to greater urinary output —› causing a decrease in blood volume and, thus, a decrease in mean cardiovascular pressure back to normal. The converse is found when astronauts return to gravity. It takes a few hours to restore normal mean cardiovascular pressure by intake of fluid and electrolytes after returning to earth, during which transition time they conserve the fluid they take in by putting out very little urine.
- During any hypovolemic shock state, such as massive hemorrhage or severe dehydration, the urinary output goes abruptly down and remains low until restoration of normal mean cardiovascular pressure.

(2) Rapid mean cardiovascular pressure buffer mechanisms:

(a) Elasticity: The elasticity of the vascular system prevents sudden blood volume loss or gain from causing a linear, temporary change in mean cardiovascular pressure. Elasticity has, of course, an instantaneous buffer effect. Evidence of this is found in one's ability to give a pint of blood at the blood bank without going into severe low cardiac output. This buffer effect bolsters circulation while the blood volume — and, thus, mean cardiovascular pressure — is restored by subsequent oral intake of fluid.

(b) Vascular/extravascular equilibrium: There is a pressure equilibrium between the various extravascular compartments of the body and the cardiovascular space (Fig. 3). Changes in mean cardiovascular pressure result in shifts of fluid back and forth which tend to buffer sudden changes. This buffer system is fairly rapid but not instantaneous, as evidenced by the observation that a person going into severe shock from sudden loss of blood would not have had the same severe shock state if the loss had occurred more gradually over a period of an hour or so.

(3) Humeral and neuro-muscular-vascular reflexes:

These responses from stimuli, which alter vascular compliance, act as buffer systems. They prevent sudden changes in mean cardiovascular pressure from sudden position changes, such as going from lying to standing. They also buffer the effect of sudden loss of blood volume from hemorrhage.

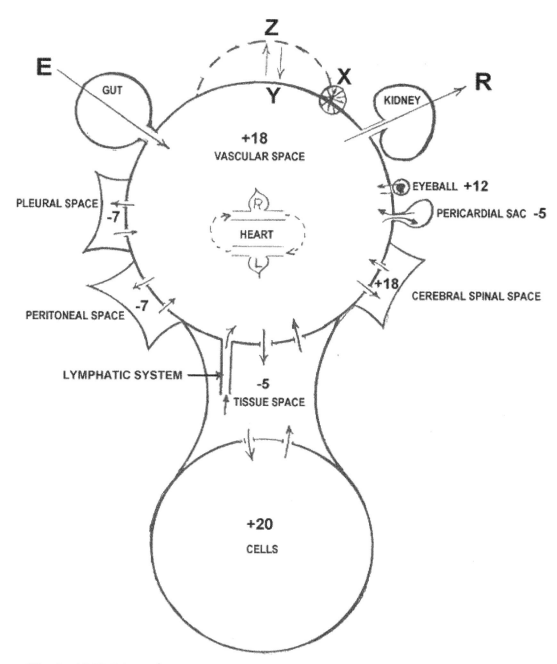

Fig.3. HYDRAULIC COMPARTMENTS OF THE BODY
All of the compartments are in pressure equilibrium with the cardiovascular system. The relative pressures result from energy forcing fluid into or out of a compartment (E) and resistance to fluid moving in the opposite direction (R). Note that the resultant pressures are negative in some spaces and positive in others. "Z" and "Y" refer to the variable elasticity of the vascular system.

Impedance to the Flow of Blood from the Outlets
Around to the Inlets of the Ventricles

Four factors tend to impede the flow of blood in the cardiovascular circle. Therefore, they are inverse determinants of cardiac output: (1) resistance, (2) elasticity, (3) limited compliance of the ventricles to filling, and (4) inertia of intermittent blood flow to the ventricles. The interrelationship of these four factors makes flow determination much more complicated than plain resistance, which is the single impediment in rigid, open ended, linear hydraulic systems.

(1) Resistance and (2) Elasticity

Unlike rigid linear hydraulic systems, where a given resistance may have the same effect on flow, irrespective of its location, the elasticity of the vascular system makes location of the resistance a significant parameter. Because of the elasticity of the circle, the location of any particular resistance determines to what extent that resistance has on impeding blood flow. A given resistance may have little or no effect in determining cardiac output if it is near the outlet of the ventricles, yet the same magnitude of resistance may have tremendous slowing effect on circulation if located near the inlets of the heart. Resistance points that have little compliant vascular bed "upstream" (arteriolar resistance), increase pump work but may not affect cardiac output significantly. The heart, except during failure, exerts enough energy to force the blood past any resistance near its outlet, with no hold up in flow. On the other hand, resistance located with a large compliant bed upstream has a tremendous effect on flow rate by slowing blood return to the ventricles. Venous sided resistance, with the compliance of the whole vascular bed upstream, is a major inhibitor of circulation rate. Thus, venous sided resistance is a major determinant of cardiac output, but arterial resistance is not. Interpolation of the additive effect of all resistance points in the circle on cardiac output must include the amount of compliant bed upstream from each resistance point.

All resistance factors — including blood viscosity, cross-section area of any vascular bed, margination of blood constituents, etc. — play roles in flow rate determination only when linked with their location to compliance upstream. Adding venous sided resistance of a magnitude that results in only a few mm. water pressure gradient may cause a marked reduction of flow. Whereas, increasing arteriolar resistance to the point of severe arterial hypertension may not appreciably change cardiac output.

In human cardiovascular physiology, the significance of resistance can only be understood when coupled with upstream compliance. It is not just the magnitude of resistance but the location of that resistance that determines its effect on circulation rate.

(3) Impediment to Ventricular Filling

The end-diastolic pressure: Ventricles not only do not suck to fill, they offer an impediment to filling. The left ventricle has thicker and stiffer walls than the right, so it tends to retard filling more than the right ventricle. Catheterization data shows end-diastolic pressure of five to ten centimeters of water above intrathoracic pressure. Don't confuse the negative pressure of inspiration, transmitted to the heart, as the heart sucking. The intracardiac pressure is always above that intrathoracic pressure (Fig. 2 at C).

(4) Inertia of Intermittent Blood Flow Offset by "The Atrial Effect"

Atrial function facilitates circulation by preventing the retarding effect that would otherwise occur from the intermittent inflow to the intermittent outflow ventricles. By being partially empty and distensible, atria prevent the interruption of venous flow to the heart that would occur during ventricular systole if the veins ended at the inlet valves of the heart.

Atria have four essential characteristics that cause them to promote continuous venous flow. (1) There are no atrial inlet valves to interrupt blood flow during atrial systole. (2) The atrial systole contractions are incomplete and thus do not contract to the extent that would block flow from the veins through the atria into the ventricles. During atrial systole, blood not only empties from the atria to the ventricles, but blood continues to flow uninterrupted from the veins right through the atria into the ventricles. (3) The atrial contractions must be gentle enough so that the force of contraction does not exert significant back pressure that would impede venous flow. (4) The "let go" of the atria must be timed so that they relax before the start of ventricular contraction, to be able to accept venous flow without interruption.

By preventing the inertia of interrupted venous flow that would otherwise occur at each ventricular systole, atria allow approximately 75% more cardiac output than would otherwise occur. The fact that atrial contraction is 15% of the amount of the succeeding ventricular ejection has led to the false conclusion that atria have their benefit by pumping up the ventricles (the so-called "atrial kick"). The real benefit is in preventing inertia and allowing uninterrupted venous flow.

The 20% to 25% increase in cardiac output from synchronized atrial function over that of atrial fibrillation doesn't belie the 75% contribution of the atrial effect, as atrial fibrillation eliminates only part of that effect. Atrial compliance, elasticity, and gravity help in emptying the atria at ventricular diastole during atrial fibrillation. Also, cardiac output during atrial dysfunction is buffered: the initial fall in circulation rate during atrial fibrillation reduces renal flow, thereby causing retention of water and the subsequent rise in mean cardiovascular pressure, which then partially offsets the slowing effect on circulation.

Thus, four factors impede the flow of blood around the cardiovascular circle: (1) resistance with (2) upstream compliance, (3) ventricular non-compliance, and (4) inertia if there is intermittent venous flow. The combined effect of these factors that impede the flow of blood to the inlet of the ventricles and, therefore, determine cardiac output in a negative way, will be referred to as: <u>inlet impedance</u>.

> **In summary, during normal physiology:**
>
> **CARDIAC OUTPUT = FUNCTION OF MEAN CARDIOVASCULAR PRESSURE / INLET IMPEDANCE**

Automatic Balancing of the Pulmonary and Systemic Blood Volumes

The passive filling characteristic of the ventricles is the feature that accounts for the automatic maintenance of blood volume equilibrium between the pulmonary and systemic vascular beds. With passively filled pumps, the relative blood volume in the two circuits is determined by their relative size and elasticity. On the other hand, if the ventricles were type #1 pumps, which actively fill, any discrepancy in the output of the two pumps, or flow through normal physiological shunts between the two vascular systems, would very quickly shift all of the blood volume into one circuit at the expense of the other, resulting in disaster.

The right and left ventricles, by filling passively, pump out whatever amount comes to them, determined by extracardiac factors. The blood volume equilibrium, therefore, is determined by the relative size and compliance of the two circuits and, to a minor extent, by the relative impediment of flow to the two pumps. It has been noted that the output of the two sides of the heart is never equal because there are physiological shunts, which connect one vascular bed to the other. The largest of these shunts, in normal physiology, are the bronchial arteries, which go from the systemic circuit to the lungs. The bronchial blood flow is a left-to-right shunt that accounts for the left ventricular output normally being at least 10% larger than the right. Because of the shunt, more blood returns to the left ventricle than the right, the left ventricle passively fills more than the right, thereby causing it to produce a greater output and thus the equilibrium is maintained. With other types of pumps this would not occur.

A dramatic illustration of this volume equilibrium, automatically being maintained during a massive discrepancy in output of the two pumps, is seen in atrial septal defects. In this case, the right ventricular output may be four or five times that of the left. Yet, the volume equilibrium is maintained. A large atrial septal defect virtually results in a single atrium above the two ventricles. The shunt occurs during ventricular diastole, because the right, thin walled ventricle is more distensible than the non-compliant thicker walled left ventricle. The blood in the common atrium goes the way of least resistance. The greater filling into the more compliant right ventricle results in greater right ventricular output. The greater right output goes to the lungs and then directly back to the right ventricle, returning again through the septal defect. Because of the passive filling, this results in no progressive increase in the blood volume in the lungs, and no disturbance in the maintenance of the blood volume equilibrium. After closure of the septal defect, resulting in much smaller right heart output, the volume equilibrium remains. This equilibrium, which persists after such sudden, massive changes in right heart output, occurs automatically because of the passive filling characteristic of the ventricles. It is the physical characteristics of the two vascular beds (e.g., relative size, compliance, and impediment to flow), that determines the volume balance with passive filling ventricles.

> **The passive filling of the ventricles accounts for the maintenance of blood volume equilibrium between the systemic and pulmonary vascular systems.**

Heart Rate and Stroke Volume

Normally, the passively filling ventricles are not maximally filled at diastole. Also, they are exerting excess energy over that needed to eject blood at the flow rate of blood entering the ventricles. With such volume and energy reserve, if blood enters the heart faster, the flow rate can go up without any change in strength of contraction or heart rate, unless the heart is being filled maximally (heart failure). Conversely, if blood enters the ventricles at a slower flow rate, the output will go down irrespective of any lowering of the rate or decrease in strength of cardiac contraction.

Thus, at a given flow rate, with the normal excess pump power, in both strength and rate, the heart rate and stroke volume become reciprocals of one another, as long as the heart doesn't go into failure. At a given flow rate, decreasing the heart rate increases the stroke volume; while increasing the heart rate decreases the stroke volume. The observation — that cardiac output usually parallels the cardiac rate and strength of contraction — has resulted in the fallacious conclusion, based on *post hoc ergo propter hoc* reasoning ("following this, therefore because of this"), that they are cause and effect.

Even though these two variables do not normally control cardiac output, their variability has physiological benefit in energy conservation. During low circulation rates, when the heart is not filling to its maximal volume during diastole, the heart down shifts its rate and strength of contraction, thus conserving energy. Also, during high output states, it increases its rate and strength of contraction, thus preventing failure from any limitation of output that might be imposed by ventricular chamber size. If the heart functioned at its maximum strength of contraction and a rate of 150 beats a minute, the cardiac output would go up and down just as it does normally. But what a waste of energy would result! So the variability of rate and strength of contraction have only ecological value: that of conserving world food supply.

Two mechanisms cause the heart to roughly parallel its energy expenditure with cardiac output, preventing failure during high output states and saving energy during low output: (1) The neural and humeral stimuli that increase mean cardiovascular pressure are also those that increase heart rate and strength of contraction. (2) Increased stretching of the ventricles at diastole causes some increased strength of contraction at systole (Starling's law of the heart). While this observation of Starling is true and contributes to the paralleling of cardiac output to strength of contraction, it is probably not a major determinant of energy expenditure, as strong cardiac contractions continue even when the heart is completely empty, as seen during cardiac bypass surgery.

The reciprocal relationship of heart rate and stroke volume can be demonstrated in patients with complete heart block by varying the rate of firing of their pacemakers. Variations in rate, within physiological levels, above that required to prevent complete diastolic ventricular filling, are associated with no change in cardiac output. Therefore, a compromise rate of 78 is a commonly used setting that allows a wide range of activity acceptable to most patients.

> **Conclusion**: During normal physiologic states, there is always pump energy excess over that used in circulation.

Exercise and Cardiac Output Increase

Two factors cause the increase in cardiac output during exercise: Exercise causes the cardiovascular impedance to be decreased and the mean cardiovascular pressure to be increased. The intermittent skeletal muscle contractions around venous beds, which contain one-way valves, act as a peripheral pump which overcomes significant impedance to flow. Neuro-humeral reflexes speed the heart rate, which slightly lowers the inlet impedance to the heart, by lowering the end-diastolic pressure over what it would otherwise be. The increased heart rate guarantees excess energy expenditure, thus preventing cardiac power failure at the higher circulation rate. The neuro-humeral response to exercise also throws the vascular system into spasm, thereby increasing the mean cardiovascular pressure from the "G-suit" effect of tensing the body in general. These are temporary changes, lasting only during the exercise period, which do not effect the long term homeostatic control of circulation.

Distribution of Blood Flow: Pulsatile Blood Flow and Arteriolar Resistance

Pulsatile arterial blood flow tends to result in diffuse, fairly equal distribution of blood to all tissues of the body. This phenomenon would not occur with a non-pulsatile steady flow. Arteriolar resistance variability from time to time and from place to place, superimposed on the otherwise diffuse distribution, controls preferential blood flow to specific areas, with physiologic benefit. The diverting of a greater portion of cardiac output to the digestive tract after meals and the increased flow to muscles during exercise are examples of changes in distribution of blood flow controlled by variable arteriolar resistance. Peripheral arteriolar resistance, rather than having a cardiac output control function, has its physiological significance by its determination of distribution of blood flow.

Chapter 2: Abnormal Circulation

Now that we have amalgamated the ten unique characteristics into an integrated concept of normal cardiovascular function, we proceed to examine the implications of those ten unique characteristics during pathological states.

Heart Failure (Pump Energy Failure) Contrasted with Pump Energy Excess

Heart failure is the only situation where the non-sucking heart determines its output. In this state, the heart is pumping at its maximum output and therefore, in effect, by limiting the output, it determines the output.

Normally, the heart rate and pump energy are in excess of that needed to eject enough blood at systole, so that the ventricles are empty enough at diastole to allow unobstructed passive filling. Normally ventricles are not maximally filled so there is reserve compliance which allows cardiac output to be determined by the extra-cardiac factors, rather than being limited by the heart. If the heart fails, it pumps less than the normally controlling extra-cardiac factors would dictate. During <u>pump energy failure</u> the heart's output is being limited by whatever amount it is able to pump. The heart in such a situation of failure, by limiting output, becomes the sole regulator of cardiac output.

<u>Definition</u>: Heart failure exists whenever heart function is inadequate to produce the output which would be allowed by the mean cardiovascular pressure and inlet impedance. Heart failure is present when cardiac function is limiting and, therefore, determining cardiac output. Normally, extra-cardiac factors being constant, if the heart contracts more strongly, efficiently, or rapidly, no increase in output occurs. Normally, the heart is exerting excess energy over that used to produce cardiac output and thus a reserve is present, so that increases in circulation rate are not impeded by the heart. Normally, at a given circulation rate, an increase in heart rate or increase in strength of cardiac contraction does not increase the output. However, during failure where the heart is pumping out the maximal circulation rate that it can produce at the moment, increases in the rate and strength of contraction do increase cardiac output, up until venous pressure is reduced to normal, at which time the heart is no longer in failure.

When heart failure occurs, blood volume equilibrium shifts to the system behind the failing ventricle with an increase in the venous pressure at the inlet of the failing ventricle. The other ventricle, in the presence of the slowed circulation rate, is able to maintain a low inlet pressure, with the resulting shift in blood volume equilibrium toward the circuit behind the failing ventricle. With left ventricular failure, for instance, the circulation slows to the rate limited by the left ventricle and a new equilibrium occurs with a shift of blood volume from the systemic circuit to the pulmonary vascular bed. Pulmonary edema occurs if this shift in equilibrium is severe.

The high end-diastolic ventricular pressure and high-end systolic volume, from a low ejection fraction in heart failure, would also be factors limiting flow, if the heart weren't already determining a lower output.

A secondary change occurs when the left ventricle fails. The lowered cardiac output decreases renal blood flow, thereby causing retention of fluid and elevation of the mean vascular pressure. Fluid equilibrium, which exists between the cardiovascular system and extra-vascular spaces, then results in edema, liver engorgement, and the other fluid abnormalities of congestive heart failure. Heart failure can intensify in a descending spiral if the blood volume progressively expands. The over-stretching of the heart puts it at a poor mechanical advantage, thus further increasing its failure (the descending limb of Starling's curve).

An extreme example of right heart failure is seen after a "Glen" right atrial-pulmonary artery anastomosis procedure for palliation of tricuspid atresia. Here, the person has virtually complete failure of a non-functioning right heart, with a surgical bypass of the right ventricle so that, in effect, the circulation is by the left ventricle alone. Circulation is slowed by the increase in impediment that normally would be overcome by the right ventricle. As a consequence of the slowed circulation, the kidneys receive slow flow, causing retention of fluid, resulting in a permanent elevation in the homeostatic level of mean cardiovascular pressure, resulting in some compensatory increase in cardiac output at the new equilibrium.

Causes of Heart Failure:

Cardiac causes (low myocardial power):

- Myocardiopathies (viral, arteriosclerotic, toxic, rheumatic)
- Incompetent leaky valves, e.g., mitral or aortic
- Abnormally slow heart rate, e.g., complete heart block

Peripheral vascular causes:

- Abnormally high mean cardiovascular pressure causing potentially higher circulation rate than a normal heart is able to deliver, e.g., renal shutdown resulting in high output failure
- Abnormally low impedance to the heart, e.g., arteriovenous fistula

Signs and Characteristics of Heart Failure:

1. High venous pressure is the major pathognomonic finding in cardiac failure. If the heart were not in failure, the pressure would be normally low.
2. A new equilibrium occurs between the two vascular systems with a shift in volume to the field behind the failing ventricle. I.e., left ventricular failure shifts the equilibrium with a greater volume in the pulmonary circuit, resulting in x-ray evidence of pulmonary vascular engorgement and high pulmonary venous pressure. With right heart failure, the neck vein pressure is seen to be elevated. However, the systemic system initially is large enough to accommodate the shift in blood volume from the lungs without significant venous engorgement until secondary changes (as seen below) occur.
3. With either left or right heart failure, increased fluid accumulation in the body progresses in a descending spiral caused by slowed cardiac output —› less than optimum renal blood flow —› low renal output —› retained fluid in the vascular system —› increased mean cardiovascular pressure —› edema —› liver engorgement and ascites.

NORMAL	HEART FAILURE
Heart energy excess	Heart energy deficit
Low ventricular inlet pressure	High inlet pressure
Incompletely filled ventricles	Completely filled ventricles
Increase in rate —› no increase in cardiac output	Increase in rate —› increase in cardiac output
Increase contraction —› no increase in cardiac output	Increase contraction —› increase in cardiac output
Cardiac output = f mcvp/inlet impedance	Heart determines cardiac output
Body water equilibrium	Progressive water retention
Normal cardiac output	Low cardiac output

Pathologic Conditions Resulting in Heart Failure:

- Complete heart block with slow pulse
- Extremely rapid pulse
- Intermittent, pulsatile venous flow
- Low ejection fraction
- Myocardiopathy with weak myocardial contraction
- Myocardial infarction to the point of low ejection volume
- Valvular heart disease with stenosis or insufficiency
- Myocardial failure
- Pericardial tamponade
- Constrictive pericarditis

Goals in the Treatment of Heart Failure:

1. Make the heart contract stronger and faster.
2. Lower the resistance to ejection of blood from the ventricles.
3. Lower the mean cardiovascular pressure by salt and water restriction and diuretics
4. Restore the atrial effect if it is compromised by atrial fibrillation or nodal rhythm.

If failure is due to myocardial contractility, lessening the failure state by lowering mean cardiovascular pressure to the point of producing normal venous pressure, without increasing myocardial energy, may create hypovolemic shock rather than restoring normal circulation.

Shock

Shock is a low cardiac output state that results from low mean cardiovascular pressure. The low mean cardiovascular pressure can result from altering of one or the other of the two determinants of that pressure: (1) Increasing vascular compliance, and (2) Loss of blood volume.

1. Low output with resulting hypotension from sudden loss of vascular tone is frequently seen after inducing spinal anesthetics. A dramatic example was seen following total central nervous system anesthesia, induced when marcaine was inadvertently injected into the spinal canal during an attempt to do an intercostal nerve block. In this case, sudden total relaxation of the entire vascular system resulted in severe hypotension. Loss of vascular tone can be reversed by the use of vasopressors and the addition of fluid.
2. Shock from loss of blood volume can be temporarily compensated for by decreasing vascular compliance by vasopressor drugs, but ultimately needs to be corrected by blood volume restoration.

A combination of the two mechanisms of shock is seen in anaphylaxis, where there is both relaxation of the vascular system plus a shift in blood volume to the extra-vascular space, caused by increased capillary porosity.

Whatever the cause, the resulting low mean cardiovascular pressure in shock can be restored to normal by rapid infusion of fluid and electrolytes, aided by the use of vasopressor drugs.

During both of the shock mechanisms, the heart is already spending excess energy, so increasing the heart rate or strength of contraction will not increase the cardiac output. The obvious increase in output that follows the use of epinephrine, or other vasopressors, in such states is not caused by the inotropic effect on the heart, but by their effect of increasing the mean cardiovascular pressure by decreasing vascular compliance. Thus another temporary therapeutic measure, which increases mean cardiovascular pressure by making the system less compliant, is the surrounding of the vascular system by a "G-suit" such as those used by emergency medical personnel. It matters little how fast or how strongly the heart contracts: If there is no pressure in the cardiovascular system, there can be no circulation.

Arterial Hypertension (Low and High Output Types)

Arterial hypertension results from two radically different mechanisms. One is caused by increased arteriolar resistance in a system with normal mean cardiovascular pressure. The other results, in the presence of normal arteriolar resistance, from elevated mean cardiovascular pressure.

1. In arterial hypertension from increased arteriolar resistance, there is no slowing of the circulation rate from the resistance. The heart, which normally expends excess energy, just forces whatever blood that comes to it right past the resistance point. The only change in circulation is the elevated arterial pressure. Such arterial abnormal resistance may have its origin from transient neuro-humeral stimuli. However, with longstanding arterial hypertension, changes eventually occur, resulting in more permanent structural resistance increase. If, eventually, the vascular changes affect blood flow to the kidneys, the second type of hypertension may develop.
2. High mean cardiovascular pressure hypertension: The most graphic example of this state is seen during renal shutdown, from whatever cause. Here, if fluid and electrolyte intake continues, and there is reduced fluid leaving the body, the vascular volume progressively rises, with a corresponding increase in mean cardiovascular pressure. If the heart is strong enough that it continues to exert excess energy over that necessary to maintain a low ventricular inlet pressure, cardiac output increases. With the increased flow, even in the presence of unaltered arteriolar state, the resulting arterial pressure is elevated. If the heart is normal, but the pressure becomes so high that venous pressure rises, high output heart failure occurs. Severe high output failure has been seen where the systemic arterial pressure was 350/200, the systemic venous pressure was elevated with the neck veins at 30 cm., the pulmonary artery pressure was doubled, and the pulmonary capillary pressure was so high as to cause pulmonary edema with frothing of blood from the mouth and nose. This is in marked contrast to the distribution hypertension caused by arteriolar resistance increase.

The spectrum of hypertension states varies greatly between these two extremes and combinations of the two. Determination of the mean cardiovascular pressure helps delineate the significance of each in diagnosis. (See the Appendix on the clinical determination of mean cardiovascular pressure.)

Therapy for the two types of hypertension should be aimed at the etiology. For high mean cardiovascular hypertension, improving renal blood flow by removing renal arterial obstruction, diuretics, low salt diet, and water restriction to lower mean cardiovascular pressure or even renal transplant may be indicated. For hypertension caused by high arteriolar resistance, arteriolar vaso-relaxing drugs and changing the patient's response to stress may be effective.

Arteriovenous Fistula

An arteriovenous fistula, with blood flow going directly from an artery to a vein without going through arterioles, capillaries, and venules, bypasses most of the resistance that has upstream compliance. Therefore, depending on size, a fistula can cause tremendous increase in cardiac output. If a shunt is great enough, high output failure can result. If a shunt is in an extremity, where it can be occluded by manual pressure, alternating decrease and increase of cardiac output can be observed, as the shunt is intermittently occluded and opened. This maneuver is a good demonstration of the peripheral vascular control of circulation rate.

Arteriovenous fistula shunts occur in a number of disease states. For example, blood flow is shunted from arteries to veins in thyrotoxic goiters. Rather than there being one large arteriovenous channel, there are many small shunting vessels. The blood may be flowing so rapidly through the many small shunts that a loud, continuous murmur may be heard over the thyroid gland. This arteriovenous shunting of blood flow accounts, in part, for the high cardiac output associated with toxic goiter.

Changes In Size of the Vascular System

Because of the peripheral control of cardiac output, changing the size of the vascular bed, permanently (such as by a leg amputation) or by short term decrease (such as by cross-clamping the aorta and vena cava during surgery), causes an instantaneous decrease in cardiac output, mechanically, without any involvement or need of neurologic or humeral reflexes.

Complete Heart Block and Fixed-Rate Pacemakers

Patients with heart block, with a very slow pulse rate, have maximal ventricular filling before the completion of diastole, thereby restricting venous flow. This restriction in venous flow causes high venous pressure and low cardiac output. As the rate is incrementally increased by the use of a pacemaker, pressure gradually falls, until the ventricles are no longer maximally filled at diastole. Further increase in pacemaker rate above that level causes no more increase in output and no further drop in the venous pressure, as ventricular capacity is no longer restrictive to flow. A rate of 60 produces this plateau in most adults at rest and in a recumbent position. If the patient is active, the pacemaker rate may need to go to 80 or even 100 before further increases in pacemaker rate cause no corresponding increase in cardiac output. When the rate is fixed at 80 and the patient is resting, the output goes down normally with a waste of energy expended by the heart, the excess not being used for circulation. The rate of 80 allows a great variation in activity without exceeding a need for pulse increase to prevent failure. If the pacemaker is set at 120, such a rate would probably accommodate cardiac output for any violent activity. However, at rest, even though the output goes down to normal, the 120 pulse tachycardia is uncomfortable, the excess myocardial energy waste is great, and myocardial oxygen availability may be taxed. Therefore, fixed pacemaker rates are usually set at about 76 to 80 as a compromise between excess myocardial energy spent for circulation at rest and needed cardiac energy for strenuous activity.

From pacemaker experience we conclude that: (1) The heart, when not in failure, is always expending excess energy over that necessary to produce cardiac output. (2) The ability to lower pulse rate below maximum during periods of low cardiac output has implications for energy conservation rather than cardiac output.

Left Ventricular Booster Pumping

Myocardial energy failure is a frequent finding during acute myocardial infarction and for short periods of time after coming off of bypass after open heart surgery. In both of these situations, left ventricular booster pumping has proved useful during a recovery period.

A competent, passive filling, pulsatile outflow, continuous inflow assist pump, placed between the left atrium and aorta will automatically lower the left atrial pressure and restore circulation rate to normal. A passive filling booster pump, which expends excess energy, put in parallel with the failing left ventricle, will make the combined output of the two normal. The combined output of two non-sucking pumps in parallel will act as one pump which produces excess to that needed for normal cardiac output. Their combined output will be controlled by the extra-cardiac factors that normally control circulation rate. Whatever amount the ventricle is unable to pump will run automatically into the booster ventricle and be ejected into the aorta. When the heart has recovered, the left atrial and arterial blood pressure will remain the same whether the pump is on or off. One method of weaning the patient off of the booster pumping is to raise the pump a few centimeters above the left atrium. Blood will then run preferentially into the heart's ventricle and be pumped out by it, if the heart is no longer in power failure. If the ventricle is still in failure, the venous pressure will rise only those few centimeters. Then whatever the heart does not pump the booster pump will.

From cardiac booster pumping, we conclude that the non-failing heart is expending excess energy to that needed at any moment to produce its output. Expenditure of energy by a parallel passive filling booster pump cannot increase the circulation rate above that dictated by extra-cardiac determinants.

SUMMARY: **The cardiovascular system is a closed elastic circle, containing two passive filling pumps in series with two vascular beds, systemic and pulmonary. Normally, the circulation rate made by the two pumps is controlled by mean cardiovascular pressure and inlet impedance. It is only during heart failure, when heart function is limiting the cardiac output, that the heart is regulating circulation rate.**

Chapter 3: Open-Heart Surgery with Passive-Filling Pumps

The advent of open-heart surgery has provided a wealth of data and observations regarding cardiovascular physiology. Furthermore, our expanding ability to correct cardiovascular defects has been a stimulus to better understand how the system works. Delineation of the essential mechanical features of the heart has allowed construction of heart replacement pumps that: (1) automatically provide normal circulation during heart bypass, with the flow rate remaining under control of normal physiological mechanisms; (2) allow research, in animals, by isolating the peripheral vascular versus cardiac effect on circulation rate from responses to various drug, humeral, and neural stimuli; and (3) provide a mechanical model having the ten unique characteristics of the cardiovascular system, for teaching and the study of circulatory phenomena.

The Mechanical Heart Replacement Pumps

The pumps have the four characteristics in common with the heart that allow them automatically, without pump regulation, to reproduce circulation under control of normal physiologic mechanisms: (1) The pumps fill passively and don't suck at their inlets. (2) The output of the pumps is pulsatile. (3) The pumps have atria that allow uninterrupted inflow to the intermittent outflow pumps. (4) The capacity of the pump ventricles is greater than any anticipated diastolic filling.

Many designs of pumps with the above characteristics could be made. One of the six variations that the author has used is shown in Figure 4. The pump consists of a flat atrial-ventricular silicone rubber tube with a reinforcing flat cotton cover (Fig. 5). The flat configuration, which prevents rebound to a round cross-section after being compressed, is responsible for the passive filling characteristic. The tube is mechanically compressed, sequentially, by four plates activated by cams and lifters (Fig. 6). Two narrow plates act as inlet and outlet valves to the ventricular portion of the tube. Two wide plates act as atrial and ventricular impellers (Fig. 4). The atrial impeller compression and sequence are critical in preventing interruption of venous inflow when the ventricles are being emptied. Therefore, atrial compression is incomplete, leaving the atria with a 3/16" channel at maximum compression. The slope of the atrial cams is gentle, to prevent any sharp rise in pump atrial pressure that might interrupt venous flow. The cam's timing is such that the "compression let go" occurs just before the inlet valve closes. (For more detailed schematics of one pump design, see Dr. Anderson's "Extracorporeal Heart" patent, listed in the "Additional Materials" at the end of the text.)

Fig. 4. A mechanical, non-sucking, continuous inflow, pulsatile-outflow pump

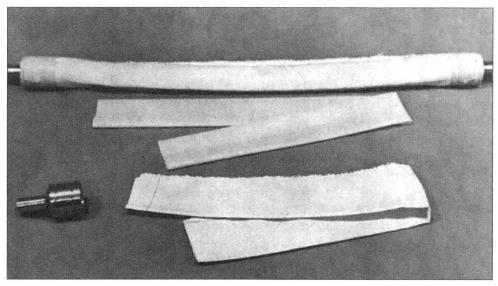

Fig. 5. The flat, non-sucking atrial-ventricular silicone rubber tube, its fabric covering, and end connectors.

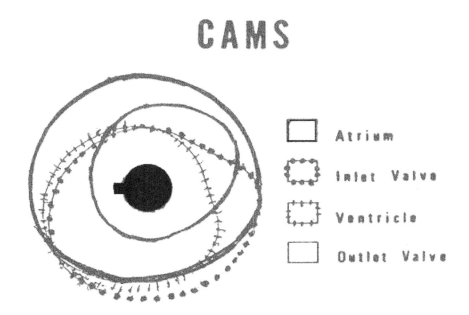

Fig. 6. The pump's cam configuration and the timing relationship of the cams.

Clinical Surgery Using the Heart-Simulating Pump

Cardiac surgery using both right and left heart bypass pumps has one significant advantage over conventional cardiopulmonary bypass: No oxygenator is used, as the lungs are left in the perfusion circuit and are provided with normal circulation. With normal pulmonary blood flow during the operation, post-operative pulmonary insufficiency is less of a problem than when the lungs are bypassed using an oxygenator.

By connecting the unique pumps in parallel to the heart's ventricles and with the pumps at heart level, cardiac function can be temporarily interrupted during surgical procedures without any interruption in normal circulation. When both the heart and bypass pumps are functioning simultaneously, the combined output remains the same as when the heart alone is pumping. Because both fill passively, the combined output remains regulated by the mean cardiovascular pressure and inlet impedance. Any amount that goes to the pumps doesn't go to the heart, and vice versa. If the heart function is then interrupted, by induced fibrillation or diversion of all the venous flow from one or both ventricles, the flow automatically goes to the pumps and circulation is uninterrupted and continues at the same rate. Subsequently, when the heart is defibrillated, or when the flow to the heart is no longer interrupted, the heart output takes over part of the circulation, with the total circulation rate still remaining the same. Weaning off of bypass is done simply by sequentially elevating the pumps a few centimeters at a time, thereby diverting more and more of the venous flow to the heart. Myocardial competence to take over the entire circulation becomes evident if the circulation rate and blood pressure remain unchanged after the incremental elevations are made. By appropriate elevation of the pumps above heart level, booster pumping can be done to maintain any desired atrial pressure in the heart during a recovery period.

The advent of open-heart surgery has provided a wealth of data and observations regarding cardiovascular physiology. Furthermore, our expanding ability to correct cardiovascular defects has been a stimulus to better understand how the system works. Delineation of the essential mechanical features of the heart has allowed construction of heart replacement pumps that: (1) automatically provide normal circulation during heart bypass, with the flow rate remaining under control of normal physiological mechanisms; (2) allow research, in animals, by isolating the peripheral vascular versus cardiac effect on circulation rate from responses to various drug, humeral, and neural stimuli; and (3) provide a mechanical model having the ten unique characteristics of the cardiovascular system, for teaching and the study of circulatory phenomena.

Specific Applications of the Mechanical Pumps

Coronary Bypass Surgery

Cardiac bypass during coronary surgery is used to provide a non-moving target with decompressed ventricles and the security that adequate circulation is being maintained during manipulation of the heart. Coronary artery surgery is not an open heart procedure, as the coronary arteries are on the surface of the heart. Therefore, closed bypass is applicable. Four cannulae are used. One pump is connected from the left atrium to the aorta, and the other from the right atrium to the pulmonary artery (Fig. 7).

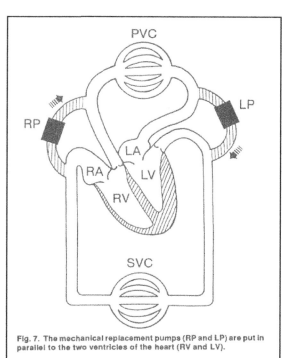

Fig. 7. The mechanical replacement pumps (RP and LP) are put in parallel to the two ventricles of the heart (RV and LV).

It is important to start the left-sided bypass before the right. If the right one is started and the heart happens to fibrillate before the left bypass is started, blood will be pumped into the pulmonary circuit while no blood is leaving it. The result would be acute pulmonary engorgement ("liver lungs") with a fatal outcome. Likewise, the left-sided bypass should be discontinued last.

The heart is electrically fibrillated while the coronary-graft anastomoses are made. Then the heart can be defibrillated before the aorta-graft anastomoses are made. If, during the coronary-graft anastomoses, greater decompression of the ventricles is desired, the atrial cannulae can be advanced into the ventricles to drain them as well as the atria. Figure 8 shows arterial blood pressures staying relatively constant before and during a coronary bypass procedure, using right and left heart bypass.

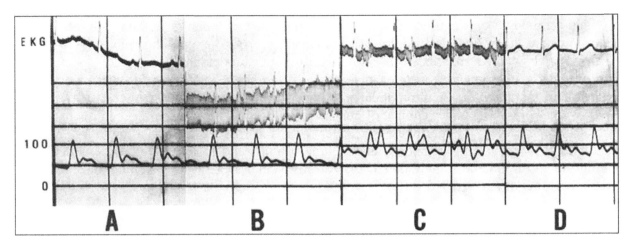

Fig. 8. Representative samples of blood pressure tracings obtained during the course of coronary bypass using the passive-filling pumps for cardiac bypass. "A" is before pump bypass, "B" is during pump bypass with the heart fibrillating, "C" is parallel pumping with the heart defibrillated, and "D" is the pressure after pump bypass.

During cardiac bypass with the two pumps, with the lungs functioning and without the use of an oxygenator, the anesthesiologist maintains ventilation and support of circulation just as he would in other non-cardiac procedures. Circulation rate reacts to blood loss, fluid infusion, and vasoactive drugs, just as when the heart is functioning.

Pulmonary Stenosis

While it might appear that pulmonary stenosis could be repaired by using a right-sided bypass alone, one laboratory experience has demonstrated this to be too hazardous for clinical application. During the course of a right heart bypass, with the left ventricle functioning, the heart unexpectedly fibrillated. With the left ventricle suddenly not pumping, and before the right heart bypass pump could be turned off and the heart defibrillated, the lungs became irreparably overloaded with blood. The resulting "liver lungs" were not reversible, causing the death of the animal. Therefore, for safety, total heart bypass is always used, even when a right heart bypass alone could allow adequate access for correction of the lesion.

Use of a Passive Filling Pump with a Bubble Oxygenator

During open-heart surgery, two pump bypass is precluded because of the difficulty in bypassing the left atrium, with its many pulmonary veins. Therefore, cardiopulmonary bypass with an oxygenator is necessary whenever the cardiac chambers need to be opened.

The passive filling, pulsatile output, continuous inflow pump has advantages over other types of pumps when used in a pump-oxygenator circuit. It can automatically produce normal circulation rate with a normal pulse wave, without any control adjustments, and without the hazard of oxygenator blood level fluctuations or air emboli.

If the inlet of the pump is placed at the priming level in the oxygenator (Fig. 9), because the pump is non-sucking, the blood will never fall below that level. Any blood that runs into the oxygenator that would tend to raise the level is automatically pumped back into the patient. The oxygenator is positioned at such a level that normal venous pressure is maintained by gravity drainage during bypass.

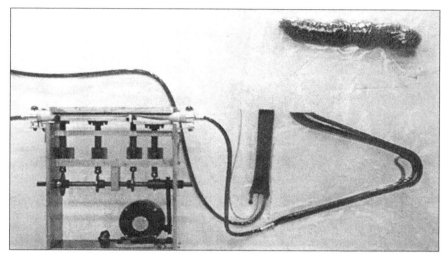

**Fig. 9, The pump used with an oxygenator.
Note that the oxygenator is primed with blood so that its outlet reservoir is at a level even with the inlet of the pump. Thus, the level cannot fall below that prime level, as the pump cannot suck. Any blood that enters that would tend to raise the level, automatically runs unimpeded into the pump.**

With this pump-oxygenator setup, circulation rate is determined by the patient's mean cardiovascular pressure and inlet impedance, just as it does in the intact body. During the bypass, circulation rate is modified by using vasoactive drugs, blood, and fluid replacement to change the mean cardiovascular pressure, not by pump alteration.

This technique — beginning partial bypass in parallel with the heart — results in normal circulation rate, as both the heart and the extra-corporeal system fill passively. When complete bypass is produced, by occluding the vena cava around the caval catheters, the circulation rate remains the same. Terminating bypass is very simple. The occluding tapes around the vena caval tubes are released and the venous drainage tubing is occluded in increments. As more and more blood is diverted to the heart, the venous pressure will indicate whether or not further decrease in extra-corporeal pumping will be tolerated. In this way, the pump-oxygenator can be used as an automatic booster device during the myocardial reovery period, following the cardiac repair procedure. Figure 10 shows the blood pressure before, during, and after bypass, including the parallel bypass at the end of the procedure, in a patient with wide-open aortic insufficiency. The wide, abnormal pulse wave of aortic insufficiency is followed with a normal pulse wave while on bypass and after valve replacement. Figure 11 shows typical pulse waves during bypass and Figure 12 shows superimposed waves during circulation produced by parallel pumping of the heart and extra-corporeal pump.

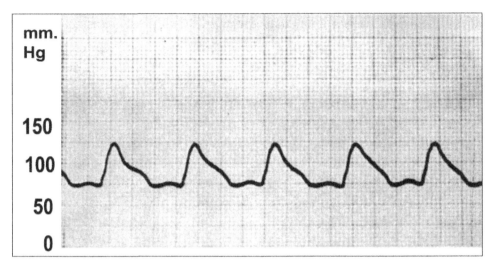

Fig. 11. Typical aortic pulse wave, obtained during pump bypass

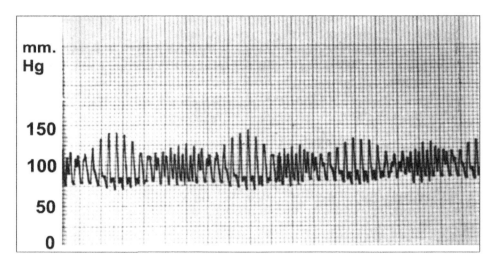

Fig. 12. Typical superimposed pulse waves of the heart and pump, functioning in parallel, with the combined output the same as with either alone.

Figure 13 illustrates a safety feature of the passive filling pump used with an oxygenator. Bypass was just underway, during a mitral valve replacement, when the inferior vena cava cannula slipped back into the right atrium. The inferior vena cava being temporarily occluded by the circumferential tape, caused the venous drainage to the pump to drop to one liter per minute. The oxygenator did not run out of blood and no air was pumped into the patient during this low output episode. After the tube was re-advanced properly into the inferior vena cava, the blood flow to and from the pump automatically returned to normal.

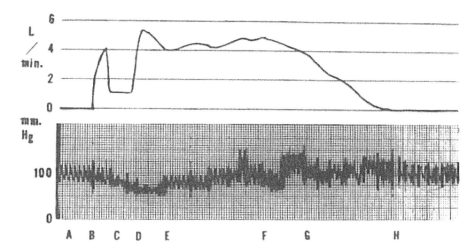

Fig. 13. The Pump flow tracing above the aortic pressure tracing demonstrates that, with the passive filling pump, the output parallels the vena-cava drainage, without the hazard of the oxygenator running out of blood if a drainage crisis occurs, as seen with other pumps.
"A-B" before pump bypass, "B-C" full bypass, "C-D" Obstruction of the inferior vena-cava flow to the pump, "D-E" restoration of full bypass, with time for re-distribution of the blood volume to a normal equilibrium between the veins and arteries, "E-F" full bypass, "F-G" Parallel pumping with heart function restored by unclamping the vena-cava tapes, "F-G" Weaning off of the pump by progressively clamping off the venous drainage tubes to the pump, "H" off bypass with normal cardiac function.

The Non-Sucking Pump Used with a Membrane Oxygenator

The non-sucking heart replacement pump is ideal for use with a membrane oxygenator. It automatically allows a constant blood volume in the system with this closed-volume oxygenator. There is no need for a reservoir for volume monitoring of changing blood levels, which would occur and require alteration of pump rate with other types of pumps. In effect, the passive filling, continuous inflow pump used with a membrane oxygenator provides a closed automatic bypass system. (*Note:* In all clinical applications of passive filling pumps, it is important to use large caliber venous drainage tubing so as not to add to the vascular system's normal inlet impedance to the pumps.)

Pulsatile Blood Flow During Bypass Surgery

Cardiac bypass during heart surgery has given an opportunity to observe the benefit of pulsatile blood flow, by comparing it with techniques using non-pulsatile flow. Pulsatile flow ensures diffuse normal distribution of blood flow to all the organs and tissue of the body, while non-pulsatile flow results in reduced flow to certain vascular beds and excessive flow to others. With non-pulsatile flow, the brain and kidneys receive reduced blood supply. Also, incomplete metabolism from islands of under-perfused tissue results in acidosis, not found when pulsatile flow is provided. When hypothermia is used, the need of pulsatile flow to insure diffuse distribution of circulation is increased. While the adverse effects of non-pulsatile flow can be offset by the use of vaso-relaxers, blood dilution, and other methods, the value of pulsatile flow in normal function is illustrated by comparing findings from the two types of pumping systems.

Ventricular Compliance — an Inlet Impedance Factor

Open-heart surgery on a patient with severe hypertrophy of the left ventricle, from longstanding aortic stenosis, illustrated how ventricular compliance can be a factor in the inlet impedance determination of cardiac output. This patient's left ventricle measured three-fourths of an inch in thickness. After replacement of the aortic valve, the bypass was discontinued. The heart contracted very strongly, with a normal pulmonary venous pressure of 12 cm. water. However, there was practically no cardiac output and the blood pressure remained at only 55/25. The blood volume was, therefore, increased by increments while watching the pulmonary venous pressure go to 16, 18, 20, 25, and 30 cm., with no effect on output or blood pressure. The heart continued to beat very forcibly. Suddenly, at a pulmonary venous pressure of 35 cm., almost at the point of causing pulmonary edema, the heart distended during diastole, cardiac output went beyond normal with a resulting blood pressure of 160/90. The non-compliance of this thick ventricle was a significant impediment to passive filling and cardiac output.

Chapter 4: Animal Experiments with Passive-Filling Pumps

Using the mechanical pumps that have four characteristics of the heart (passive filling, pulsatile outflow, uninterrupted inflow, excess flow capacity) to replace heart function in dogs, it is possible to demonstrate the significance of those characteristics in determining the following circulation phenomena:

1. Blood volume balancing between the systemic and pulmonary circuits

2. Blood volume's effect on mean cardiovascular pressure and thus on cardiac output

3. Circulation rate control by the peripheral vascular effects of chemical, humeral, and neural stimuli

4. The atrial effect in providing uninterrupted venous flow to the heart

5. Pulsatile blood flow causing diffuse distribution of circulation

The Dog-Pump Model

The following demonstrations were obtained during cardiac replacement of both right and left ventricles with passive filling pumps in greyhound dogs. One pump was hooked up with a cannula connected from the right atrium to the pulmonary artery; the other was connected from the left atrium to the aorta (Fig. 7). After the pumps were started, the hearts were fibrillated with a low voltage alternating electric current, thereby diverting all of the blood flow to the pumps. The pump rate was set arbitrarily at 80 beats per minute. The ventricles had a maximum capacity of 80 cc. (Figs. 14 & 15).

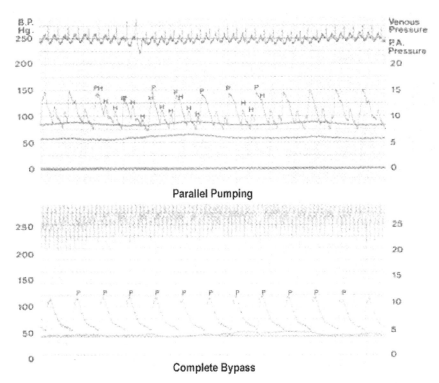

Fig. 14. Arterial blood pressure in a greyhound dog with the heart replacement pumps in parallel with the heart. The graph shows parallel pumping, "H" is the cardiac ejection, "P" is the pump ejection and "PH" is a fusion beat.

Fig. 15. Arterial pressure on complete right and left heart bypass with the heart fibrillating.

Blood Volume Balancing Between the Pulmonary and Systemic Circuits

The model with non-sucking pumps gives an excellent opportunity to demonstrate the automatic blood volume balancing between the two circuits by passive filling pumps. The balancing can be demonstrated by putting shunts between the two circuits, in various locations, that would cause imbalance with other types of pumps.

Figure 16 shows a biopsy of normal appearing lung, obtained after bypassing the heart with the non-sucking pumps for a short time. Figure 17 shows the same lung after 2 hours, where a shunt of 1 liter per minute — a total of 160 liters of blood — was shunted from the aorta to the pulmonary artery. The shunt was a 1/4 inch tube placed between the two arteries. An adjustable roller pump, incorporated in the tubing, produced the shunt. During this time, without any adjustment, the left heart replacement pump automatically put out approximately one liter more blood per minute than the right pump, thereby maintaining the blood volume balance between the two circuits. The function of the lungs remained normal in the presence of great disparity in circulation rate between the systemic and pulmonary circuits.

Fig. 16. Lung structure before right and left heart bypass with two passive filling, mechanical pumps.

Fig. 17. The lung after two hours of bypass, including a variety of shunts, totalling 160 liters of blood shunted from the systemic circuit to the lungs.

Contrasted with this experiment was one where the heart was replaced with two roller pumps, adjusted so that the output of each was identical. When a 1 liter per minute right-to-left shunt was placed in the circuit, within 15 seconds the lungs became so engorged and hemorrhagic that they bled from the bronchi and appeared beefy red and consolidated like liver (Fig. 18 and Fig. 19).

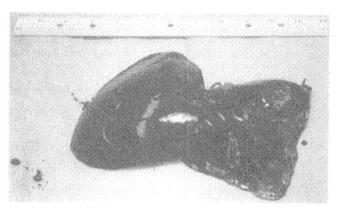

Fig. 19. "Liver lung" caused by one liter imbalance in pulmonary-systemic blood flow.

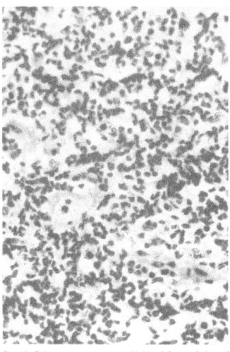

Fig. 18. Pulmonary engorgement with blood, from unbalanced blood flow from use of roller pumps with a shunt of one liter to the pulmonary circuit.

With the passive filling pumps, reversing the great vessel shunt and introducing shunts at other locations all failed to create a disequilibrium between the blood volumes in the systemic and pulmonary circuits.

Figure 20 is a record of the effect of adding fluid to the systemic venous system on blood flow from the pumps. The two lower superimposed tracings are the flow rates of the right and left passive filling pumps. At the beginning of the graph, the flow to the pulmonary artery is slightly more than the flow to the aorta. Shortly thereafter, the pulmonary artery flow is slightly less. At point A of Figure 20, 200 cc. of blood is rapidly injected into a systemic vein. At point B, the right pump puts out more volume than the left, adding the injected fluid to the previous blood flow. When the bolus of fluid reaches the left pump (Fig. 20, from B to C), the right passive filling pump automatically has a greater output. At point C, the increased blood volume caused by the fluid addition is redistributed back to a normal equilibrium, with the pulmonary and systemic vascular beds each having received their representative volume expansion by the transfused fluid.

This is a demonstration of how intravenous infusions and transfusions, even though introduced into only one circuit, do not cause an imbalance in the blood volumes between the lungs and the rest of the body. The new equilibrium occurs automatically, without any sensing devices or neuro-chemical feedback mechanisms, as the non-sucking pumps have none. With passive filling pumps, it makes no difference where the fluid is introduced, no disequilibrium can occur.

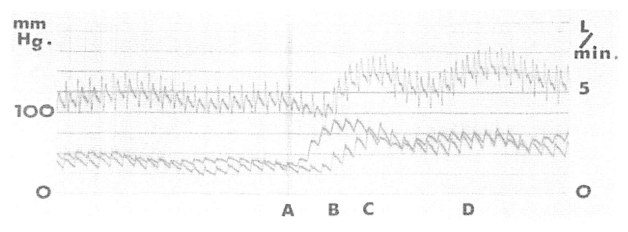

Fig. 20. Simultaneous, superimposed right and left pump flow rates.
From "A" to "B" 200 cc. of blood was infused into a systemic vein. At "B" the pulmonary artery flow is greater than the aortic. At "C" the flow from the pumps returns to near equality, after the new blood volume equilibrium between the lungs and systemic circuit has occurred.

Single Ventricle Replacement

The right ventricle was bypassed with a pump from the right atrium to the pulmonary artery. The pulmonary artery was clamped between the pulmonary valve and the pulmonary artery cannula, thereby diverting all of returning systemic blood to the pump.

With the right ventricle replaced, circulation continued in a normal fashion, with the systemic-pulmonary blood equilibrium staying unaltered. The lung's function and appearance remained normal in all but one case.

A complication occurred the second time this experiment was done on a dog, which prevented the disaster that might have resulted if it had been used in a human heart surgery case. The single right heart bypass, with the left heart still functioning, was anticipated to be used in cases of pulmonary valvular stenosis. Five minutes into that experiment, the heart developed ventricular fibrillation; the pump didn't. The fibrillating ventricle of course quit its pumping, so no blood left the lungs thereafter. The pump, in the next three to four beats, overloaded the lungs with enough blood to give them the appearance of liver. The result was total loss of pulmonary function with no chance of recovery of the animal.

This experience not only prevented the use of this procedure in human heart surgery, but gave one other significant benefit: The exposed hazard made it clear what the order of cannulation should be in two ventricle bypass. The left heart should be cannulated first and the pump started. If the heart should happen to fibrillate before the right cannulation was completed, pumping all of the pulmonary blood into the systemic circuit would not be disastrous because of the lung's comparative low blood volume. Then the right cannulation can be done, and total bypass then initiated.

The Relationship of Blood Volume to Mean Cardiovascular Pressure and Cardiac Output

Figure 21 is a baseline graph obtained from a dog with his heart function replaced by two passive filling pumps. His vascular volume is normal. The pressures were taken without the presence of any vascular stimulants or depressants, except for a pentothal anesthetic. In order to determine the mean cardiovascular pressure, the pumps were turned off for 30 seconds. In 10 seconds the venous pressure had risen to 16 cm. of water pressure, and the arterial pressure had dropped from 125/55 to 16 cm. as well. This pressure — 16 cm. of water — that equalized in the entire pump and vascular system is the mean cardiovascular pressure. The mean cardiovascular pressure can be read more accurately on the venous recording than the arterial because of the difference in the scale.

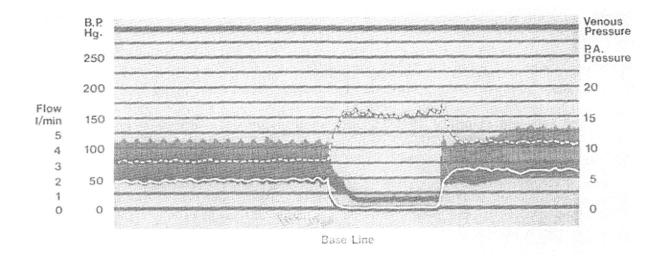

Fig. 21. Approximate normal baseline pressures, during mechanical heart replacement with non-sucking, pulsatile outflow, continuous inflow pumps in a dog.

Dashed line -- Venous pressure
                                                                           8 cms. water
Solid white line --- Pump flow rate
                                                                           2 liters / min.
Arterial pressure -- 120/50
Mean-cardiovascular pressure ------------------------------------ 16 cm. water

Hypovolemia

Low blood volume in a dog, which had moderate hypertension, was produced by draining 600 cc. of blood from the venous cannula. Figure 22 was taken after withdrawal of the 600 cc. This produced a drop in pump output from the control of 2 liters per minute to less than 1 liter per minute, and a drop in the mean cardiovascular pressure from 16 cm. water pressure to approximately 10 cm., and a drop in arterial pressure to 120/50 from an original pressure of 160/90. Figure 23 was obtained after restoring the original blood volume by transfusion of the 600 cc. of blood. The mean cardiovascular pressure went back to the original 16 cm., and the flow rate returned to 2 liters per minute. Note that the arterial blood pressure overshot the original pressure. It then took several minutes to come back to normal. These findings parallel those seen in the intact animal, thus illustrating the relationship of mean cardiovascular pressure, resulting from blood volume and cardiovascular compliance, to cardiac output.

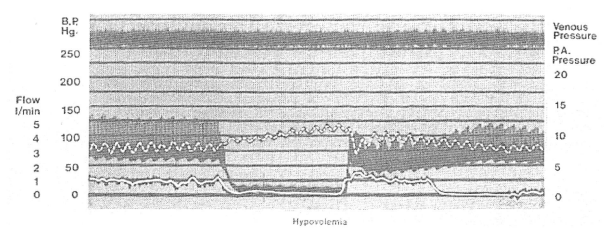

Fig. 22. Hypovolemia causing low mean-cardiovascular pressure and the resulting low pump output.

Dashed white line ———————————————————————— Venous pressure
                                                    7 cms. water
Solid white line —————————————————————————— Pump flow rate
                                                    < 1000 cc. / min.
Arterial blood pressure ——————————————————— 120/50
Mean-cardiovascular pressure ———————————— 11 cms. water

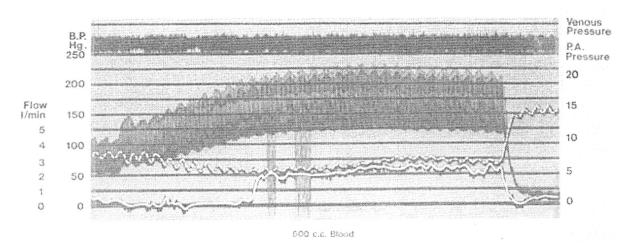

Fig. 23. Restoration of mean-cardiovascular pressure by transfusion of 600 cc. of blood, resulting in restoration of normal pump output and arterial blood pressure.

Dashed white line ———————————————————————— Venous pressure
                                                    6 cms. water
Solid white line —————————————————————————— Pump flow rate
                                                    2500 cc. / min.
Mean cardiovascular pressure ————————————— 16 cms. water

The Effects of Peripheral Vascular Response from Chemical Stimuli on Cardiac Output

With a large number of stimuli that can affect either cardiac or peripheral-vascular funtion, or both, it is important to know which effector site is responsible for the resulting circulation change. *Post hoc ergo propter hoc* reasoning ("following this, therefore because of this") has misled thinking in the past. One example of an erroneous conclusion by this reasoning is that variations in heart rate from stimuli are responsible for variations in cardiac output. Heart rate, myocardial contractility, ejection fraction, onset to peak time, end diastolic pressure, arteriolar resistance, and venous pressure are all responses to stimuli from baroreceptors, neurologic stimuli, humoral vascular stimuli, arteriolar vasopressors*, vasorelaxers*, and stretch receptors. The pump-animal preparation allows visualization of cause and effect of many cardiovascular phenomena in a way not possible in the intact animal. Knowing that the mechanical pump cannot react to neurological or chemical stimuli, we can separate changes that are due to a peripheral vascular response from those that are due to a cardiac response. If a stimulus causes the same response as when the heart is in the circuit, we know that the response is from the peripheral vascular effect alone. Then, conversely, any change found with the heart functioning, which is absent with the pumps in the vascular system, can be interpolated as a change that results from a cardiac response. Thus, the animal-pump model allows us to demonstrate whether the effect of a specific stimulus on circulation is due to a cardiac or a peripheral vascular response.

*Note: The use of terms "vasoconstrictors" and "vasodilators" are conventionally used for stimuli that would better be referred to as "vasopressors" and "vasorelaxers," as blood is non-compressible. For every increment that vessels constrict in one place, there must be a simultaneous dilation somewhere else in the vascular system. When one segment dilates, there must be a corresponding constriction elsewhere, unless additional fluid enters the vascular space. Unless referring to a localized area of the vascular bed, such as the arterioles, it is more accurate to refer to vasopressors and vasorelaxers.

Vasopressors

Figures 24 and 25 document responses to epinephrine, with the heart function replaced with the passive filling pumps. Figure 24 shows an increase in the circulation rate going from 4 1/2 liters per minute to 7 liters per minute after intravenous injection of 1 cc. of 1/10,000 epinephrine. Figure 25 shows the mean vascular pressure, from a normal of 15 cm. water, going to 19 cm. after the epinephrine. Any slowing effect on pump output that might be expected because of increased resistance from spasm of the vascular system is obviously offset by the increased circulation rate caused by the increase in mean cardiovascular pressure. The increase in arterial blood pressure, in this case, is from a combination of increased arteriolar resistance and elevation in pump output from the increased mean cardiovascular pressure.

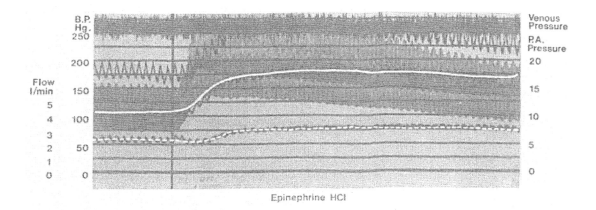

Fig. 24. Response to epinephrine
Broken white line ————————————————————— Venous pressure
Solid line ————————————————————————— Pump output
Arterial blood pressure ———————————————— 150/75
Vertical black line ————————————————— Epinephrine injected

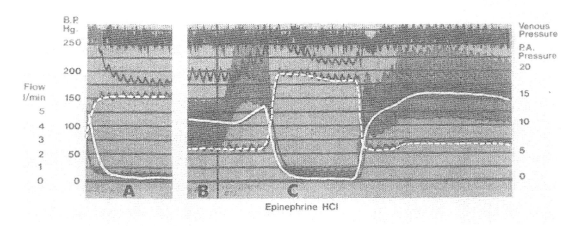

Fig. 25. Mean cardiovascular pressure after I.V. Epinephrine
Broken white line ————————————————————— Venous pressure
Solid white line ————————————————————— Pump output
Vertical black line ————————————————— Epinephrine injected
"A" Before epinephrine, Pump turned off
    Mean-cardiovascular pressure ——————————— 15 mm. water
"B" Before epinephrine, Pump turned on
Vertical black line ————————————————— Epinephrine injected
"C" After epinephrine, Pump turned off
    Mean-cardiovascular pressure ——————————— 18 mm. water
Notice the elevation of mean-cardiovascular pressure increase to 18 mm. water, with the blood pressure response.

Figures 26 and 27 show the circulation response to neo-synephrine in an animal preparation, with the heart replaced with a passive filling pump. The pump has no control or regulating devices. The circulatory changes are, therefore, without any cardiac effect and are purely from the effect of the drug on the vascular system. Note that the increased arteriolar resistance, having little upstream compliance, does not decrease the flow rate. The increased mean cardiovascular pressure, which increases the circulatory rate, more than offsets the outflow impediment of the arteriolar resistance. Note also that the pulmonary artery pressure rises as well as the systemic pressure in response to neo-synephrine. The venous pressure stays the same after the injection, showing that the pump is not in failure at the increased flow rate.

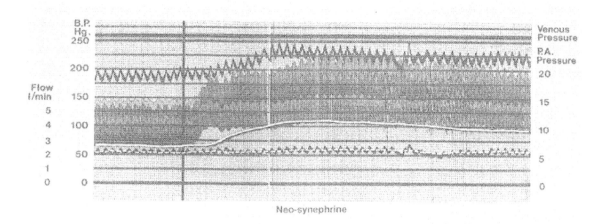

Fig. 26. After neo-synephrine injection the blood pressure, in this normally hypertensive greyhound dog, goes very high. The pump output increases from 3 to 4 1/2 liters / minute.
The circulatory response, without the heart in the circuit, is the same as normal with the passive filling mechanical non-regulated pumps.
Dashed white line ---Venous pressure
Solid white line --Pump flow rate
Vertical black line ---Neo-synephrine injected

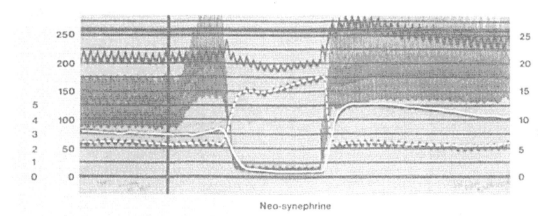

Fig. 27. Thirty seconds after injecting neo-synephrine, when the circulatory response is evident, the pumps were turned off. The resulting mean-cardiovascular pressure had increased from a previously recorded level of 13 cms., in this animal (not shown), to 17 cms. water pressure. Circulation increased from 3 to 5 liters / min. from the vaso-pressor.
Dashed white line ---Venous pressure
Solid white line --Pump flow rate
Vertical black line ---Neo-synephrine injected
Mean-cardiovascular pressure ---17 cms. water pressure

Dopamine (Figures 28 and 29) and Levophed (Fig. 30) cause the same circulation response with the pulsatile outflow, uninterrupted inflow, passive filling mechanical pump as when the heart is in the circuit. This finding supports the thesis that vasopressors alter circulation by their peripheral vascular effect, and that the cardiac response to the drugs has its effect in preventing heart failure during variation in circulation rates.

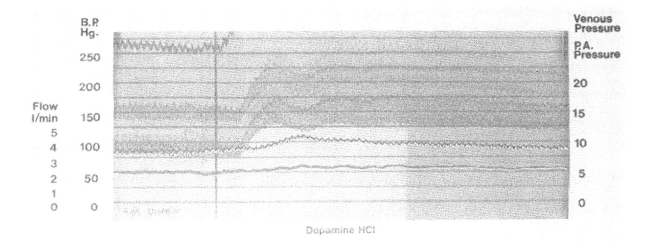

Fig. 28. The peripheral vascular response to Dopamine.
Dashed white line —————————————————————————————Venous pressure
Solid white line ——————————————————————————————Pump flow rate
Vertical black line —————————————————————————————Dopamine Injection
The arterial blood pressure response far exceeded the modest increase in pump output.

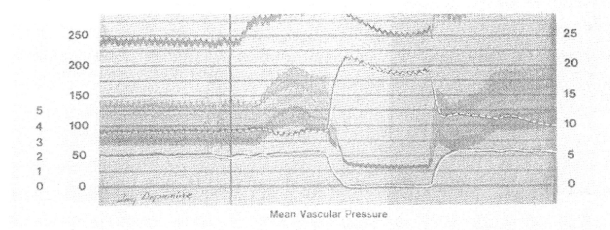

Fig. 29. Mean-cardiovascular pressure response to Dopamine.
Pump output increases 250 cc. / min. with a mean-cardiovascular pressure of 20 cm. water.

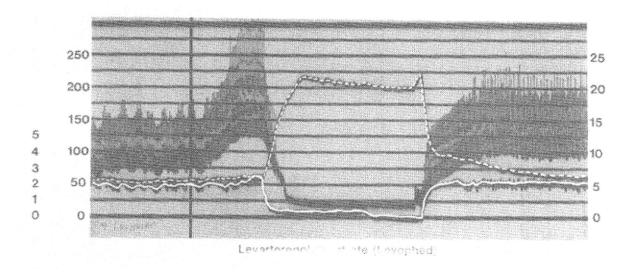

Fig. 30. Circulation response to Levophed with passive filling pump-animal preparation.

Vaso-Relaxers

Figures 31 and 32 show responses from injection of nitroprusside intravenously in a dog whose heart function is replaced with the passive filling pumps. Figure 31 shows the reduction in arterial blood pressure and a reduction in circulation rate similar to that seen in the intact animal.

Figure 32 shows the circulation response to nitroprusside with cardiac function replaced by passive filling pumps. Twenty seconds after the nitroprusside injection took effect in lowering the arterial pressure, the pump was turned off. The venous pressure rose and the arterial pressure fell to the mean cardiovascular pressure of 8 to 10 cm. water (normal being 16 to 18 cm. water). The circulation rate change was inconsistent, sometimes falling slightly and sometimes staying about the same. This occurs as the nitroprusside decreases inlet impedance to the extent that it offsets any circulation rate decrease that otherwise would occur from lowering mean cardiovascular pressure alone.

Isoproterinol (Fig. 33) causes the same drop in blood pressure and slight increase in circulation rate as in the intact animal.

Other vasorelaxer drug effects are illustrated by Figures 34, 35, and 36.

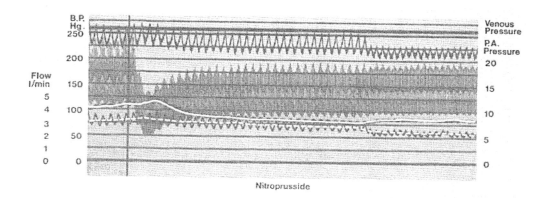

Fig. 31. Nitroprusside decreases the arterial blood pressure by its peripheral vascular effect, with passive filling pumps replacing the heart. The vertical line indicates the time of injection of nitroprusside.
Dashed white line ———————————————————————————— Venous pressure
Solid white line ——————————————————————————————— Pump output
Vertical black line ————————————————————————————— Nitroprusside injection

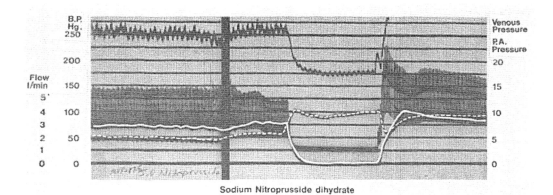

Fig. 32. Nitroprusside response on circulation
Dashed line ———————————————————————————————— Venous pressure.
Solid white line ——————————————————————————————— Pump output.
Mean-cardiovascular pressure ———————————————————— 10 cm. water.
Vertical black line ————————————————————————————— Nitroprusside injection

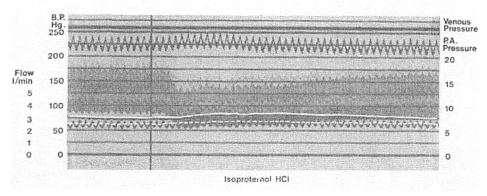

Fig. 33. The circulatory response to Isuprel.

Chapter 4: Animal Experiments with Passive-Filling Pumps

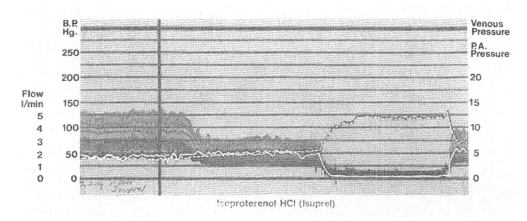

Fig. 34. Isuprel drops the blood pressure, but has little effect on pump output, even in the pressence of decreased mean-cardiovascular pressure to 12 cm. water.

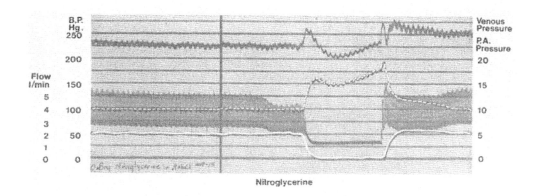

Fig. 35. Nitroglycerine peripheral vascular effect on circulation.

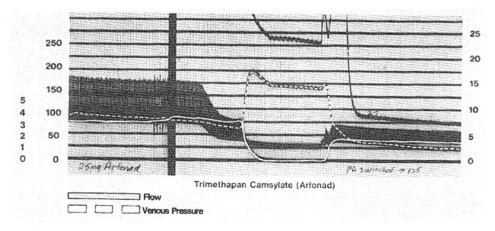

Fig. 36. Arfonad has a marked effect in lowering both arterial blood pressure and pump output by relaxing the vascular system thereby decreasing the mean-cardiovascular pressure.

Conclusion

The vasopressor and vasorelaxer findings illustrate that the circulation changes in response to these stimuli are the result of their peripheral vascular effect. Any cardiac change in rate or strength of contraction in the intact animal from those stimuli is responsible for the heart staying out of failure concomitant with the peripheral vascular circulatory rate control changes. The pump-animal preparation becomes a useful study method of separating the cardiac from the peripheral vascular effect of drugs on circulation.

The Atrial Effect Experiment

The atrial portion of the pump was eliminated by blocking the atrial impeller in the down position, thereby rendering the atrium a fixed diameter tube, with its cross-section the same as that of the vena cava drainage tubing. In effect, this modification created a passive filling pump with the veins going directly to the ventricles without an intervening atrium. It was then possible to abruptly go back and forth from a pumping circuit, with or without an atrium, while making no other change. At a normal mean cardiovascular pressure and a pump rate of 80, with no atrial effect, and the inertia of stopped venous flow at each pump beat, the pump output was only 1/4 of that seen with the atria functioning. In the absence of atria, when the pump rate was progressively increased, the output was further decreased. At a pump rate of 120, which provided normal flow with the atria functioning, the circulation stopped completely. At this rate, whenever the stopped venous blood would start to flow again toward the ventricles, the inlet valve would close again so no flow would occur. With no atrial function, progressively lowering the flow rate below 80 caused some increase in flow over the 25%, as there was enough time between beats for the venous flow to accelerate toward the ventricles. At a rate of 30 or below, the atria provided no output benefit, as there was plenty of time for the interrupted venous blood to accelerate toward the ventricles. However, at that slow rate the ventricles were in failure (limiting flow by being maximally filled at end diastole). The atrial effect is the major factor in lessening the inlet impedance to passive filling pumps. The atrial effect is most important at a pump rate in the range of normal heart function.

Pulsatile Blood Flow Experiment

In order to observe circulation with and without pulsatile flow, the omentum of a dog was placed on the stage of a dissection microscope. Two parallel tubes, joined at the end with a "Y" connector, were interposed in the omental artery. Blood to the omentum could be diverted through one tube or the other by cross-clamping one of them. One of the tubes contained an elastic section with a slightly constricted outlet, which would damp out the pulse wave and make uninterrupted flow to the omentum when the other tube was clamped. Using this setup, the omental circulation could be observed during either pulsatile or non-pulsatile flow. During pulsatile flow, blood flow remained unchanged and stable through the many vessels. With non-pulsatile flow at the same rate, a marked change in distribution of blood flow progressively occurred. More and more of the flow went to fewer and fewer channels. After 15 minutes, a few vessels had dilated with most of the blood going through them, while other vessels had become empty with no flow. There were small islands of omentum with no blood flow next to others with higher than normal flow. When pulsatile flow was re-initiated, it took two minutes for the original flow distribution to be established. Parenthetically, the easiest way to produce a diffuse hydraulic distribution system is to make it pulsatile like the vascular system.

Chapter 5:
Hydraulic Model of the Cardiovascular System

A model, incorporating many of the unique characteristics of the cardiovascular system, allows the opportunity to observe the significance of each of those features. The characteristics reproduced in the model are: (1) the hydraulic system is a circle; (2) it is an elastic system, (3) it is filled with fluid producing a mean pressure; (4) there are two pumps in series between two vascular beds; (5) the pumps fill passively; (6) pump output is intermittent; (7) the pumps have atria that allow continuous, unimpeded flow to the pumps; (8) the pumps are capable of pumping out a greater volume than the system will produce; and (9) there are resistance points near the pumps' inlets as well as near their outlets.

Studying the model has an advantage over studying *in vivo* preparations in that the effect of altering one variable at a time can be directly observed. Furthermore, the model eliminates the possibility that some simultaneous, hidden, undetected change might have occurred that would invalidate cause and effect conclusions (Fig. 37). The model has proven to be a very effective study and teaching tool.

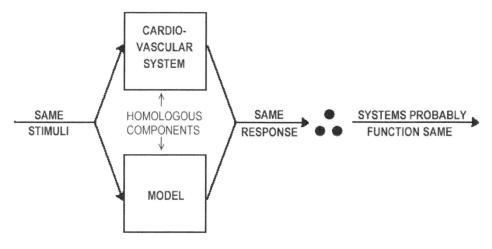

Figure 37. Rationale

The Model:

Figure 38 shows the layout of the model pictured in Figure 39.

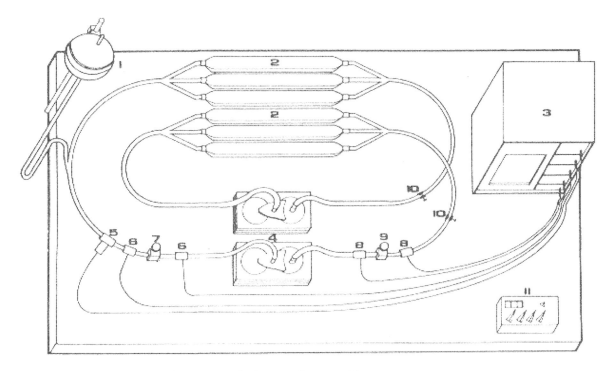

Fig. 38. Circulation model—schematic.
1—Reservoir for adding or subtracting "mean-cardiovascular-pressure".
2—Elastic systemic and pulmonary vascular beds.
3—Flow meter and pressure recorder.
4—Right and left passive filling pumps.
5—Flow probe.
6—Inflow pressure transducers.
7—Inflow resistance adjustment clamp.
8—Outlet pressure transducers.
9—Outlet resistance adjustment clamp.
10—Resistance clamps, movable to any site.
11—Ventricular rate and atrial timing controls.

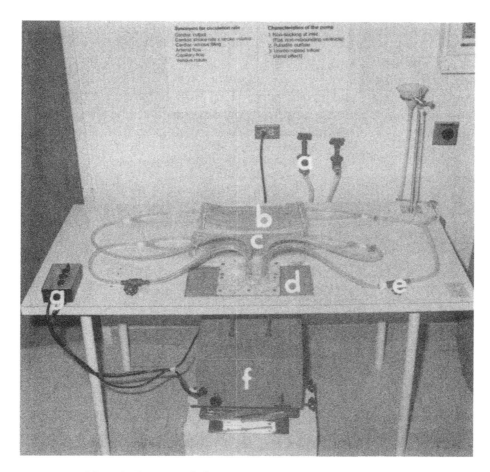

Fig. 39. Circulation model.
   a—Positive and negative air pressure sources.
   b—Systemic vascular bed.
   c—Pulmonary vascular bed.
   d—Right and left heart pumps.
   e—Metering valve.
   f—Solenoid valves to positive and negative pressure tanks.
   g—Ventricular rate and atrial timing controls.

The Pumps

The model pumps share the three unique characteristics of the heart: They are non-sucking, and therefore fill passively at their inlets, they have a pulsatile outflow, and they have atria which allow uninterrupted inflow to the intermittent, pulsatile outflow ventricles.

Figure 40 shows one of the two assembled pumps. Figure 41 shows the unassembled parts of a pump.

The atrial and ventricular pump chambers consist of a rigid side (Fig. 41 at A) and an opposing pliable silicone rubber side (Fig. 41 at B). The ventricular half of the rigid side has both an inlet and an outlet port containing inlet and outlet valves, respectively. The atrial half of the rigid side has a single port with no valve. There is a "Y" connection between the atrial port, inlet port to the ventricle and the venous line. This "Y" inlet allows venous fluid to run both into the atrium, when the inlet ventricular valve is closed during systole, and into the ventricular chamber during both ventricular diastole and atrial systole. This arrangement reproduces the "atrial effect" by allowing continuous uninterrupted flow from the veins to the intermittent outflow ventricles, thus preventing the need of overcoming inertia by starting stopped flow after each pump beat.

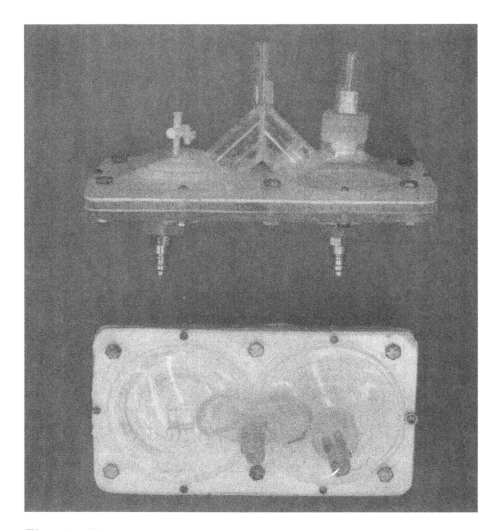

Fig. 40. The passive filling pump.

The impeller part of the pump also consists of a rigid side (Fig. 41 at E) and an opposing pliable rubber side (Fig. 41 at D). Each half of the rigid part has a port through which air can be forced into or sucked out of the impeller chambers. When the impeller chambers are inflated, their rubber side exerts pressure on the silicone rubber side of the ventricle or atrial chambers, thus emptying the ventricle or atrium.

A spacer (Fig. 41 at C), placed between the opposing impeller and pumping chambers, has many side holes which allow air to pass freely to and fro between the ambient air and the space between the impeller and pump ventricles. This free communication prevents any negative pressure, caused by suction used to deflate the impeller spaces, from being transmitted to the pump ventricles. Thus, the ventricles fill passively from venous pressure.

The power supply to the pumps is from compressed air and vacuum sources, delivered alternately to the impeller spaces timed in the desired sequence. An electronic regulator triggers solenoid valves to control pulse rate and the ratio of systole to diastole. The pulse rate is variable from 1 to 160 strokes per minute at an impeller pressure of one to thirty p.s.i. Reducer valves regulate peak pressure and onset to peak time. Impeller deflating suction is strong enough that passive ventricular filling is not impeded.

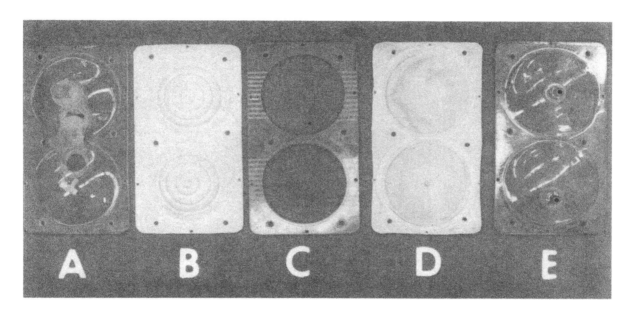

Fig. 41. The components of the pump. "A and E" are the outer rigid shell halves, "B and D" are the silicone rubber impeller diaphragms, "C" is the spacer with holes to the ambient air.

Simulated Vascular Network

The major vessels are silicone tubing, while the capillary beds are made of more compliant Penrose drain material. Variable resistance points are at various sites in both arterial and venous lines. Pressure transducers connect to a direct writing recorder and pressure tubes connect to an electromagnetic flow meter. A transfusion reservoir provides opportunity to change the fluid volume within the system. Different weights placed on a plate resting on the "capillary beds" allow changes in the system compliance to be made. The system was filled with water. Hundreds of pieces of #0 silk suture material, 2 millimeters in length (found to be isobaric with water), were suspended in the water to allow visualization of the circulation.

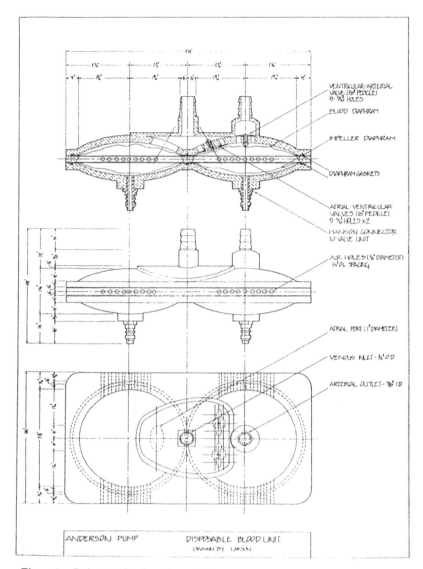

Fig. 42. Schematic drawing of the pump.

Findings Correlate with Observations in Human Physiology

Experiment 1: The Relationship of Mean Cardiovascular Pressure to System Volume

With the pumps turned off, the compliance remaining unchanged, and starting with the system full of water, at a volume of 2000 cc. at a pressure of zero, pressure was recorded as fluid was added by way of the transfusion reservoir. Figure 43 shows the non-linear relationship between the mean system pressure and the volume.

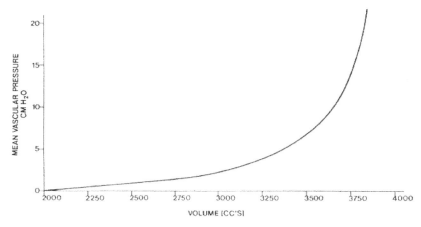

Fig. 43. Volume-pressure compliance of the simulated vascular system.

Experiment 2: The Relationship of Circulation Rate to Mean System Pressure

Starting with a mean system pressure of zero, and with arbitrary settings of the resistance points, the pumps were started. Appropriate amounts of fluid were added to increase the mean system pressure by increments of 2 cm. water pressure (Fig. 44)

Findings:

1. Figure 44 shows a direct correlation between pump output and mean system pressure over the range of 0 to 20 cm. of water pressure. The greater the mean system pressure, the greater the pumps' output.
2. At no time, over the range of pressure studied, did the ventricles fill to capacity at diastole. There was always a "pump capacity excess."
3. At no time was there any interruption of venous flow by the intermittent closing of the inlet valves of the ventricles at systole, as the atria had been emptied during ventricular diastole.

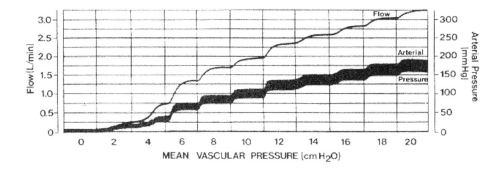

Fig. 44. The correlation of flow rate and mean-system-pressure.

Experiment 3: Pump Rate Correlation with Pump Output

The pump rate in the model was varied from 0 to 170 beats per minute with other variables remaining constant (Fig. 45). The mean system pressure was kept at 12 cm. water pressure, with the impeller pressure at 14 p.s.i. and the resistance kept constant.

Findings:

1. At pump rates between 0 and 20/min.:
    - The pump output correlated in a linear way with the pump rate (Fig. 45, from A to halfway to B).
    - The ventricles filled to capacity at each ventricular diastole.
    - The atria filled to capacity during each ventricular systole.
2. At pump rates between 20 and 40/min.:
    - The pump rate correlated in a non-linear way with the pump output (Fig. 45, from halfway in between A and B, to B).
    - The ventricles filled slightly less than capacity at ventricular diastole.
    - The atria were filled to capacity at the end of ventricular systole.
    - Venous flow was slowed but not completely interrupted before the end of ventricular systole.
3. Increasing the pump rate from 50 to 110:
    - Caused no increase in pump output (Fig. 45, B to C).
    - Was associated with progressive decrease in ventricular end-diastolic volume.
    - Caused progressive decrease in atrial diastolic volume.
    - Was unaccompanied by any interruption in venous flow.
    - The pump rate and stroke volume were reciprocals of one another.
4. Increasing the pump rate from 110 to 170 beats/min. (Fig. 45, C to D) was accompanied by a progressive increase in residual air pressure in the ventricular impeller after ventricular systole which:
    - Progressively decreased pump output.
    - Impeded ventricular filling at ventricular diastole.
    - Caused venous flow interruption at ventricular systole.

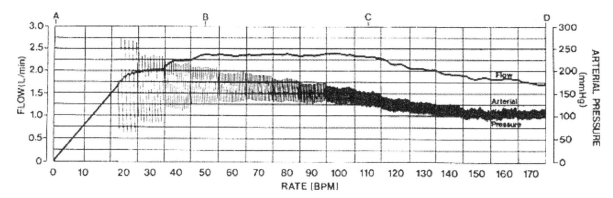

Fig. 45. Relationship of flow rate to pump rate.

Experiment 4: Ventricular Impeller-Force Correlation with Pump Output

The ventricular impeller pressure was varied from 0 to 20 p.s.i., in increments of 2 p.s.i., while keeping other variables constant (Fig. 46). The pump rate was 80, the mean system pressure was 12 cm. water pressure and all resistances remained unchanged.

Findings:

1. At impeller pressures between 0 and 10 p.s.i. there was:
   - Linear correlation between impeller pressure and pump output (Fig. 46, from A to B).
   - The ventricles were never completely emptied at ventricular systole.
   - The ventricles were completely filled at ventricular diastole.
   - Venous flow to the pumps was interrupted at each ventricular systole.
2. At impeller pressures between 10 and 18 p.s.i. there was:
   - No increase in the output as the pressure increased (Fig. 46, from B to C).
   - Maximal ventricular emptying at systole.
   - Sub-maximal filling of the ventricles at diastole
   - No venous flow interruption at any time.
3. As the impeller pressure was progressively increased above 18 p.s.i. (Fig. 46, from C to D) there was progressively incomplete evacuation of compressed air in the ventricle during diastole which:
   - Progressively decreased pump output.
   - Impeded ventricular filling at diastole.
   - Caused venous flow interruption at ventricular systole.

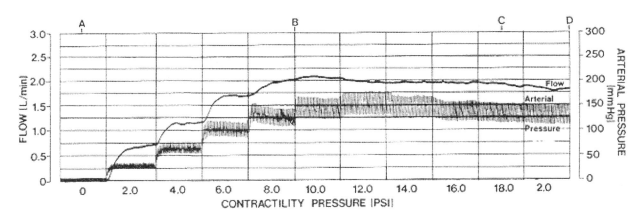

Fig. 46. Correlation of pump power to pump output.

Experiment 5: Relationship of Resistance to Pump Output

Resistance at a variety of sites was changed in the model's vascular network while maintaining other factors constant, with the mean system pressure at 12 cm. water pressure, pump rate at 80 beats per minute, and an impeller pressure of 16 p.s.i. There was a marked difference in response to a given resistance depending on where the resistance was placed in the model. Responses fell into two categories which became more obvious the closer the resistance was to either the inlet or outlet of a pump. Therefore, resistance was studied in two situations: near a pump inlet, where there was a large compliant bed upstream; and near a pump outlet, with no compliant bed upstream.

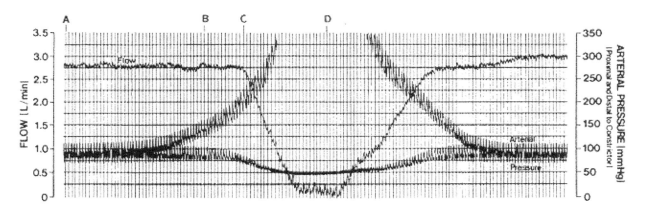

Fig. 47. Pump outflow resistance correlation with pump output.

Relationship of Resistance Near a Pump Outlet to Pump Output: (Fig. 47)

A variable resistance clamp was placed 20 cm. from the left heart homologue pump outlet (Fig. 38, #9), across which a pressure gradient could be monitored by transducers (Fig. 38, #8).

Findings:

1. As the resistance is progressively increased (Fig. 38, #9), the superimposed arterial pressure tracings, which demonstrated no gradient initially (Fig. 47 at A), separated as resistance was added (Fig. 47, from A to C).
2. With increase in outflow resistance up to 100 mm. Hg. gradient, there was no drop in pump output (Fig. 47 at C).
3. Further increase in gradient from 100 to 250 mm. Hg did cause a corresponding drop in flow (Fig. 47, from C to D).
4. In the range where there was no drop in flow (Fig. 47, A to C), the left ventricle did not fill maximally, but did empty maximally, and there was no interruption of venous flow at any time during the pumping cycle.
5. However, when the gradient became great enough that the flow slowed (Fig. 47, C to D), the left ventricle progressively emptied less completely, and the residual volume resulted in complete left ventricular filling at diastole and interruption of venous inflow at each systole.

6. Whenever the left ventricle was being filled completely at diastole, increasing the mean system pressure caused no concomitant increase in pump output.
7. As slowing of flow occurred (Fig. 47, C to D), there was gradual increase in a greater than previous volume in the pulmonary circuit at the expense of the systemic circuit. The greater the slowing, the more the volume equilibrium shifted to the simulated pulmonary circuit.

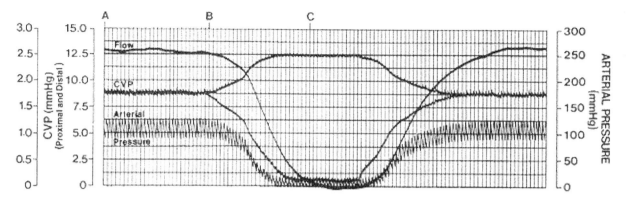

Fig. 48. The correlation of pump output with inlet resistance.

The Effect of Resistance Near a Pump Inlet on Pump Output: (Fig. 48)

An adjustable resistance clamp was placed 20 cm. from the left pump's inlet (Fig. 38, #7), across which a pressure gradient could be monitored using the transducers (Fig. 38, #6). The clamp was progressively closed. The resulting inflow resistance magnitude is represented by the pressure gradient between the transducers. Figure 48 shows the two superimposed pressure lines (Fig. 48, A to B) in the absence of resistance. When resistance is progressively increased, starting at B, a pressure gradient becomes evident (Fig. 48, B to C).

Findings:

1. There was a direct inverse correlation between pump output and inflow resistance to the pump (Fig. 48, B to C) over the whole range studied. This is in marked contrast to outflow resistance of the previous experiment (Fig. 47, B to C), where decrease in flow did not occur until there was pump failure.
2. As inlet resistance increased there was progressively less filling of the ventricles at diastole. Maximal emptying did occur at each ventricular systole, and venous flow remained uninterrupted at all times.
3. Progressive increase in left pump inflow resistance shifted the fluid equilibrium between the pulmonary and systemic circuits toward the pulmonary bed.
4. The circuit with the greatest inflow resistance was the major determinant of the output of both.

Experiment 6: The Atrial Effect

In the model, to eliminate the atrial effect, the air pressure in the atria can be left continuously obliterating the atrial chambers, while not impinging on the connection of the venous tubes to the ventricles. By intermittently producing circulation with and without the atrial effect, its contribution to pump output was determined. At pump rates between 60 and 90 there was four times the pump output (2400 cc./min.) with the atria functioning as with no atrial effect (600 cc./min.). The output without atria progressively decreased as the rate was increased. At 130 beats per minute, flow stopped completely. It is demonstrated that atria markedly increase flow when used with pulsatile, passively filling pumps. They prevent inertia, which would otherwise occur if the inlet valves were allowed to stop venous flow at each beat.

Experiment 7: Pulsatile Flow

A branching hydraulic system was made with twenty, 3/16 inch diameter, transparent tubes, six to twelve inches in length, linked together with "Y" connectors in the configuration of a simulated vascular bed.

Findings:

1. With pulsatile flow there was flow in all parts of the system. It fairly danced with the pulsation. There were areas where flow was in one direction during most of the cycle and reversed in the other direction for the rest of it. With pulsatile flow, the distribution remained stable for a long period of time.
2. With continuous non-pulsatile flow, the distribution started out in a diffuse manner. Very shortly, some channels had higher flow than others. Yet there was some flow in all of them. However, within a short time the 2 mm. silk particles began clustering at "Y" junctions, eventually totally blocking some and partially blocking others. Before long, almost all of the flow was in just a few channels and very little was in the others. When pulsatile flow then replaced the steady flow, the aggregates of silk were broken up, the whole bed began dancing with the pulse, and flow was again established in a diffuse manner, with flow fairly equal in all channels.

Conclusion:

There is obviously a difference in the mechanism of aggregation of silk particles in this system and the mechanism that caused poor distribution with non-pulsatile flow in the animal in the previous chapter. However, this experiment does show that the simplest way to guarantee diffuse, equal distribution of fluid in a hydraulic system is to make flow pulsatile.

The Model As a Teaching Tool

An example of the teaching and learning benefit of the model is illustrated by an incident. One day I had scheduled a replacement of an aortic and mitral valve in a woman who had severe mitral and aortic stenosis. The student on my service came to me and said, "This woman is so critically ill that she may have difficulty surviving such a big operation. Maybe it would be the better part of valor to correct only one valve. Then, when she is better, go back and replace the other one." I asked him which valve he thought I should replace first. He responded that he would go to the model and put an inlet and outlet obstruction on the circuit. Then remove one and then the other and see what improvement would result from a single obstruction removal. When he only removed the inlet obstruction, the increased flow to the pump put it into power failure. When he removed only the outlet obstruction it decreased the workload of the pump, but the output didn't increase.

We would have sealed the woman's fate if we had corrected only one of the valves. She would have remained in low output failure if we had corrected the aortic valve alone; and she would have gone into acute left ventricular failure if we had relieved the mitral stenosis alone. Therefore, we replaced both valves and the patient made an excellent recovery.

> **The close correlation of model findings to physiological observations in man makes the model useful in understanding how the cardiovascular system works, and helps to anticipate cause and effect in human physiology.**

Summary

The significance of ten unique characteristics of the cardiovascular system has been demonstrated by clinical observations, cardio-bypass data, animal mechanical heart replacement experiments, and a simulated cardiovascular model. All of the evidence validates that:

- The cardiovascular system is a circle containing a mean cardiovascular pressure dependent upon volume and compliance.

- The heart is a passively filling pump.

- Atria increase cardiac output by causing continuous venous flow to the intermittently contracting ventricles, not by pumping up the ventricles.

- Circulation rate is normally determined by the extra-cardiac factors: Mean cardiovascular pressure and inlet impedance.

- It is only during heart failure that the heart is the regulator of cardiac output.

Appendix: Clinical Determination of Mean Cardiovascular Pressure

A useful approximation of the mean cardiovascular pressure can be made without the need of stopping the heart and letting the pressure equalize in the entire cardiovascular system. The assumption is made that the same pressure would be found by isolating a representative sample of the arteries, capillaries, and veins at ventricular diastole, and letting the pressure equilibrate in that sample. By instantaneously interrupting arterial inflow to the arm and venous outflow from it, the pressure will fall in the arteries and rise in the veins until they are equal. This equalized pressure, which will occur within 30 seconds, approximates that found when circulation is stopped in the entire body.

Equipment:

A narrow pneumatic blood pressure cuff (approximately one inch in width) is used so that during inflation it will not displace any blood volume distally into the arm.

A one liter air pressure reservoir, pressurized to 300 mm. Hg, is connected to the pressure cuff with a valve interposed to allow instantaneous inflation of the cuff. The reservoir is pressurized by a regular blood pressure inflating bulb and valve, interposed with a "Y" connecter between the reservoir and the valve.

Two pressure transducers and recorders are needed, to allow pressure recording without any loss of fluid from the vascular bed, which would occur with a simple manometer system.

Technique:

1. The patient should be lying perfectly horizontal.
2. Both an artery and a vein are cannulated at the antecubital area of the arm and connected to the pressure transducers.
3. The blood pressure cuff is applied above the biceps bulge of the arm. It must be applied loose enough that it does not cause any venous obstruction, as evidenced by the observation that it produces no elevation of venous pressure above that seen before its application.
4. The cuff must be applied tight enough that its inflation, to 300 mm. Hg pressure, completely interrupts the arterial flow to the arm. Complete interruption can be assumed if arterial and venous pressure approach equalization after 30 seconds, and do not continue to rise thereafter.

5. The arm should be abducted sufficiently from the side of the body, so as to obtain the lowest venous pressure reading, thereby being certain that any abduction is not mechanically interfering with venous flow in the axilla.
6. The arm should be horizontal, with the anterior surface of the antecubital skin at mid-chest position.
7. The thickness of the chest should be accurately measured. The pressures are small, so careful standardization is necessary in order to get meaningful data.
8. The patient is instructed to leave the arm perfectly relaxed, and not to contract any muscle.
9. The valve or switch in the pressure line is turned, thereby inflating the cuff, which simultaneously interrupts venous and arterial flow to and from the arm.

Pressure readings are made 30 seconds after the occlusion. As the equalization process progresses, the pressure gradient becomes so small between the arteries and veins that the pressures may never completely equalize. However, the average of the arterial and venous pressures recorded at 30 seconds can be interpreted as the mean cardiovascular pressure.

Establishing such a standardization makes these approximate readings clinically useful. In any low arterial blood pressure situation, the mean cardiovascular pressure indicates whether the problem is from low blood volume or myocardial failure. mean cardiovascular pressure is a diagnostic tool that can separate high output hypertension from the more common arteriolar resistance hypertension, thus indicating the proper therapy regimen.

About the Author

Robert M. Anderson, M.D. (1920-2010), was Associate Professor of Surgery and Associate Dean of the University of Arizona College of Medicine, as well as Chief of Cardiothoracic Surgery at University Medical Center in Tucson, Arizona. As a Fellow of both the American College of Cardiology and the American College of Surgeons, and a Diplomate of the American Board of Surgeons, Dr. Anderson devoted most of his academic career to cardiovascular research and teaching.

A pioneer in heart surgery and an inventor of cardiopulmonary bypass equipment, Dr. Anderson was the author of more than sixty scientific papers. His mechanical heart patent (U.S. Patent no. 3,518,033, applied for Nov. 28, 1966) entitled "Extracorporeal Heart," was based on physiologic factors that his research delineated as essential for mechanical heart replacement. The mechanical model of the cardiovascular system presented in this book incorporates those factors. The model proved to be such an effective teaching tool that Dr. Anderson hoped others would reproduce it to introduce their students to cardiovascular physiology.

Originally published in 1993, the printed version of *The Gross Physiology of the Cardiovascular System* (ISBN 0961752815) is currently available at medical school libraries throughout the United States. A video demonstrating the concepts in this book, entitled "The Determinants of Cardiac Output," is available online at YouTube and Vimeo.com, and is available in the University of Arizona Health Sciences Library (library call number WG 106 VT 2 1980).

Additional Resources by Dr. Anderson

Anderson, Robert M. "The Determinants of Cardiac Output." Tucson, AZ: University of Arizona Department of Biomedical Communications, 1980. Videocassette, 22 min.
- Available in the University of Arizona Health Sciences Library (http://sabio.library.arizona.edu/record=b5633790~S1)
- Available online at YouTube: http://www.youtube.com/watch?v=LftmowNNAbc
- Available online at Vimeo: http://vimeo.com/6752103

Anderson, Robert M. "Intrinsic Blood Pressure." *Circulation* **9 (1954): 641-647.**
- Available online at http://circ.ahajournals.org/content/9/5/641.full.pdf

Anderson, R.M., O'Hare, J.E., and Fritz, J.M. "Automatic Demand Pump for Cardiopulmonary Bypass." *Annals of Thoracic Surgery* **10 (1970): 424-431.**
- Available online at http://ats.ctsnetjournals.org/cgi/reprint/10/5/424.pdf

Anderson, Robert M. "Extracorporeal Heart." United States Patent no. 3, 518,033 (June 30, 1970).
- Available online at http://www.google.com/patents/US3518033.pdf

Anderson, R.M., Fritz, J.M., and O'Hare, J.E. "The Mechanical Nature of the Heart as a Pump." *American Heart Journal* **73 (1967):92-105.**

Anderson, R.M., Larson, D.F., and Lundell, D.C. "The Interrelationship of Factors Controlling Cardiac Output." *Medical Hypothesis* **10 (1983): 77-95.**

Please visit
http://www.cardiac-output.info
for additional, downloadable versions of this text, and other materials.

Made in the USA
Middletown, DE
02 June 2025

76438836R00044